CW00371465

Engineering Science 2
Checkbook

J O Bird
BSc(Hons), AFIMA, TEng(CEI), MITE

A J C May
BA, CEng, MIMechE, FITE, MBIM

Butterworth Scientific
London Boston Durban Singapore Sydney Toronto Wellington

First published 1982

© Butterworth & Co (Publishers) Ltd 1982

British Library Cataloguing in Publication Data

Bird, J.O.
  Engineering science 2 checkbook. — (Butterworths
  technical and scientific checkbooks)
  1. Engineering
  I. Title   II. May, A.J.C.
  620   TA145

  ISBN 0-408-00691-9
  ISBN 0-408-00627-7 Pbk

Typeset by Scribe Design, Gillingham, Kent
Printed in Scotland by Thomson Litho Ltd., East Kilbride

Contents

Note to Reader

As textbooks become more expensive, authors are often asked to reduce the number of worked and unworked problems, examples and case studies. This may reduce costs, but it can be at the expense of practical work which gives point to the theory.

Checkbooks if anything lean the other way. They let problem-solving establish and exemplify the theory contained in technician syllabuses. The Checkbook reader can gain *real* understanding through seeing problems solved and through solving problems himself.

Checkbooks do not supplant fuller textbooks, but rather supplement them with an alternative emphasis and an ample provision of worked and unworked problems. The brief outline of essential data—definitions, formulae, laws, regulations, codes of practice, standards, conventions, procedures, etc—will be a useful introduction to a course and a valuable aid to revision. Short-answer and multi-choice problems are a valuable feature of many Checkbooks, together with conventional problems and answers.

Checkbook authors are carefully selected. Most are experienced and successful technical writers; all are experts in their own subjects; but a more important qualification still is their ability to demonstrate and teach the solution of problems in their particular branch of technology, mathematics or science.

Authors, General Editors and Publishers are partners in this major low-priced series whose essence is captured by the Checkbook symbol of a question or problem 'checked' by a tick for correct solution.

Preface

This textbook of worked problems provides coverage of the Technician Education Council level 2 unit in Engineering Science (syllabus U80/734, formerly U76/053). However it can be regarded as a basic textbook in engineering science for a much wider range of courses, such as equivalent Australian TAFE courses.

The aims of the book are, in the first place, to give a basic mechanical and electrical science background for engineering technicians, and secondly and more specifically, to develop the student's understanding of d.c. and a.c. electrical circuits, electromagnetism, statics, dynamics, energy and machines.

Each topic considered in the text is presented in a way that assumes in the reader only that knowledge attained in Physical Science 1 (syllabus U80/682) and Mathematics 1 (syllabus U80/683). This practical Engineering Science book contains over one hundred and eighty illustrations, nearly two hundred detailed worked problems, over six hundred further problems with answers.

The authors would like to express their appreciation for the friendly co-operation and helpful advice given to them by the publishers. Thanks are due to Mrs Elaine Woolley for the excellent typing of the manuscript. Finally the authors would like to add a word of thanks to their wives, Elizabeth and Juliet, for their continued patience, help and encouragement during the preparation of this book.

J O Bird
A J C May
Highbury College of Technology
Portsmouth

Butterworths Technical and Scientific Checkbooks

General Editors for Science, Engineering and Mathematics titles:
J.O. Bird and A.J.C. May, Highbury College of Technology, Portsmouth.

General Editor for Building, Civil Engineering, Surveying and Architectural titles:
Colin R. Bassett, lately of Guildford County College of Technology.

A comprehensive range of Checkbooks will be available to cover the major syllabus areas of the TEC, SCOTEC and similar examining authorities. A comprehensive list is given below and classified according to levels.

Level 1 (Red covers)
Mathematics
Physical Science
Physics
Construction Drawing
Construction Technology
Microelectronic Systems
Engineering Drawing
Workshop Processes & Materials

Level 2 (Blue covers)
Mathematics
Chemistry
Physics
Building Science and Materials
Construction Technology
Electrical & Electronic Applications
Electrical & Electronic Principles
Electronics
Microelectronic Systems
Engineering Drawing
Engineering Science
Manufacturing Technology
Digital Techniques
Motor Vehicle Science

Level 3 (Yellow covers)
Mathematics
Chemistry
Building Measurement
Construction Technology
Environmental Science
Electrical Principles
Electronics
Microelectronic Systems
Electrical Science
Mechanical Science
Engineering Mathematics & Science
Engineering Science
Engineering Design
Manufacturing Technology
Motor Vehicle Science
Light Current Applications

Level 4 (Green covers)
Mathematics
Building Law
Building Services & Equipment
Construction Technology
Construction Site Studies
Concrete Technology
Economics for the Construction Industry
Geotechnics
Engineering Instrumentation & Control

Level 5
Building Services & Equipment
Construction Technology
Manufacturing Technology

1 Simple d.c. circuits

A. MAIN POINTS CONCERNED WITH SIMPLE D.C. CIRCUITS

1   (i) All substances are made from **elements** and the smallest particle to which an element can be reduced is called an **atom**.

   (ii) An atom consists of **electrons** which can be considered to be orbiting around a central **nucleus** containing **protons** and **neutrons**.

  (iii) An electron possesses a **negative charge**, a proton a **positive charge** and a neutron has **no charge**.

  (iv) There is a force of **attraction** between oppositely charged bodies and a force of **repulsion** between similarly charged bodies.

   (v) The **force** between two charged bodies depends on the amount of charge on the bodies and their distance apart.

  (vi) **Conductors** are materials that have electrons that are loosely connected to the nucleus and can easily move through the material from one atom to another. **Insulators** are materials whose electrons are held firmly to their nucleus.

  (vii) A drift of electrons in the same direction constitutes an **electric current**. Thus electric current is the rate of movement of charge in a circuit. The unit of current is the ampere (A).

2 The unit of **charge** is the **coulomb** (C), where one coulomb is one ampere second. (1 coulomb = 6.24×10^{18} electrons.) The coulomb is defined as the quantity of electricity which flows past a given point in an electric circuit when a current of one ampere is maintained for one second. Thus, charge in coulombs, $Q = It$, where I is the current in amperes and t is the time in seconds.

3 The unit of **force** is the **newton** (N), where one newton is one kilogram metre per second squared. The newton is defined as the force which, when applied to a mass of one kilogram, gives it an acceleration of one metre per second squared.

4 The unit of **work** or **energy** is the **joule** (J), where one joule is one newton metre. The joule is defined as the work done or energy transferred when a force of one newton is exerted through a distance of one metre in the direction of the force. Energy is the capacity for doing work.

5 The unit of **power** is the **watt** (W), where one watt is one joule per second. Power is defined as the rate of doing work or transferring energy. Thus, power in watts,

$$P = \frac{W}{t},$$

where W is the work done or energy transferred in joules and t is the time in seconds. Hence energy in joules, $W = Pt$.

6 Although the unit of work or energy is the joule, when dealing with large amounts of work or energy, the unit used is the **kilowatt hour (kW h)**, where

1 kW h = 1000 watt hours
      = 1000 × 3600 watt seconds or joules
      = 3 600 000 J

7 The unit of **electric potential** is the **volt (V)**, where one volt is one joule per coulomb. One volt is defined as the difference in potential between two points in a conductor which, when carrying a current of one ampere dissipates a power of one watt, i.e.

$$\text{Volts} = \frac{\text{watts}}{\text{amperes}} = \frac{\text{joules/second}}{\text{amperes}} = \frac{\text{joules}}{\text{ampere seconds}} = \frac{\text{joules}}{\text{coulombs}}$$

A change in electric potential between two points in an electric circuit is called a potential difference. The **electromotive force (e.m.f.)** provided by a source of energy such as a battery or a generator is measured in volts.

8 The unit of **electric resistance** is the **ohm (Ω)**, where one ohm is one volt per ampere. It is defined as the resistance between two points in a conductor when a constant electric potential of one volt applied at the two points produces a current flow of one ampere in the conductor.

Thus, resistance in ohms $R = \dfrac{V}{I}$,

where V is the potential difference across the two points in volts and I is the current flowing between the two points in amperes.

9 The reciprocal of resistance is called **conductance** and is measured in siemens (S).

Thus conductance in siemens, $G = \dfrac{1}{R}$,

where R is the resistance in ohms.

10 When a direct current of I amperes is flowing in an electric circuit of resistance R ohms and the voltage across the circuit is V volts, then

Power in watts, $P = VI = I^2 R = \dfrac{V^2}{R}$

11 Terms, units and their symbols

Quantity	Quantity symbol	Unit	Unit symbol
Length	l	metre	m
Mass	m	kilogram	kg
Time	t	second	s
Velocity	v	metres per second	m/s or m s^{-1}
Acceleration	a	metres per second squared	m/s^2 or m s^{-2}
Force	F	newton	N
Electrical charge or quantity	Q	coulomb	C
Electrical current	I	ampere	A
Resistance	R	ohm	Ω
Conductance	G	siemen	S
Electromotive force	E	volt	V
Potential difference	V	volt	V
Work	W	joule	J
Energy	E (or W)	joule	J
Power	P	watt	W

12 Multiples and submultiples

Prefix	Name	Meaning
G	giga	multiply by 1 000 000 000 (i.e. 10^9)
M	mega	multiply by 1 000 000 (i.e. $\times 10^6$)
k	kilo	multiply by 1000 (i.e. $\times 10^3$)
m	milli	divide by 1000 (i.e. $\times 10^{-3}$)
μ	micro	divide by 1 000 000 (i.e. $\times 10^{-6}$)
n	nano	divide by 1 000 000 000 (i.e. $\times 10^{-9}$)
p	pico	divide by 1 000 000 000 000 (i.e. $\times 10^{-12}$)

13 **Ohm's law** may be stated as: The current flowing in a circuit is directly proportional to the applied voltage, and inversely proportional to the resistance.

$$I = \frac{V}{R} \text{ or } V = IR \text{ or } R = \frac{V}{I}$$

14 For resistors connected in series, the equivalent resistance R_T is given by

$$R_T = R_1 + R_2 + R_3 + \ldots\ldots\ldots + R_n, \text{ for } n \text{ resistors.}$$

15 For resistors connected in parallel the equivalent resistance R_T is given by

$$\frac{1}{R_T} = \frac{1}{R_1} + \frac{1}{R_2} + \frac{1}{R_3} + \ldots\ldots\ldots + \frac{1}{R_n}, \text{ for } n \text{ resistors.}$$

16 For the special case of two resistors connected in parallel,

$$R_T = \frac{R_1 R_2}{R_1 + R_2} \quad \text{(i.e. } \frac{\text{product}}{\text{sum}}\text{)}$$

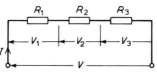

Fig 1

17 For the series circuit shown in *Fig 1*:
(a) $V = V_1 + V_2 + V_3$
(b) $V_1 = IR_1, V_2 = IR_2, V_3 = IR_3$
(c) $R_T = R_1 + R_2 + R_3$

18 The voltage distribution for the circuit shown in *Figure 2(a)* is given by:

$$V_1 = \left(\frac{R_1}{R_1 + R_2}\right) V$$

$$V_2 = \left(\frac{R_2}{R_1 + R_2}\right) V$$

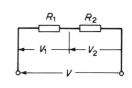

The circuit shown in *Fig 2(b)* is often referred to as a **potential divider** circuit. Such a circuit can consist of a number of similar elements in series connected across a voltage source, voltages being taken from connections between the elements. Frequently the potential divider consists of two resistors as shown in *Fig 2(b)*, where

$$V_{OUT} = \left(\frac{R_2}{R_1 + R_2}\right) V_{IN}$$

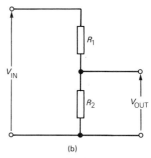

(b)

Fig 2

3

19 When a continuously variable voltage is required from a fixed supply a single resistor with a sliding contact is used. Such a device is known as a **potentiometer** (see Chapter 5, page 58).

20 For the parallel circuit shown in *Fig 3*:

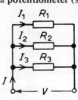

(a) $I = I_1 + I_2 + I_3$

(b) $I_1 = \dfrac{V}{R_1}$, $I_2 = \dfrac{V}{R_2}$, $I_3 = \dfrac{V}{R_3}$

(c) $\dfrac{1}{R_T} = \dfrac{1}{R_1} + \dfrac{1}{R_2} + \dfrac{1}{R_3}$

Fig 3

21 The current division for the circuit shown in *Fig 4* is given by:

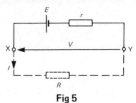

$$I_1 = \left(\frac{R_2}{R_1 + R_2} \right) I$$

$$I_2 = \left(\frac{R_1}{R_1 + R_2} \right) I$$

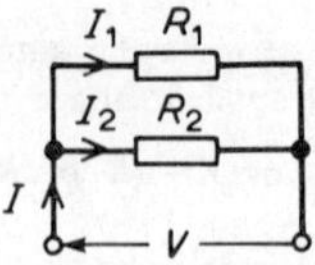

Fig 4

22 (i) The **electromotive force (e.m.f.)** E, of a cell is the p.d. between its terminals when it is not connected to a load (i.e. the cell is on 'no-load').

(ii) The voltage available at the terminals of a cell falls when a load is connected. This is caused by the **internal resistance** of the cell which is the opposition of the material of the cell to the flow of current. The internal resistance acts in series with other resistances in the circuit. *Fig 5* shows a cell of e.m.f. E volts and internal resistance r. XY represents the terminals of the cell. When a load (shown

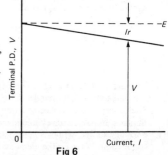

Fig 5

as resistance R) is not connected, no current flows and the terminal p.d., $V = E$. When R is connected a current I flows which causes a voltage drop in the cell, given by Ir. The p.d. available at the cell terminals is less than the e.m.f. of the cell and is given by $V = E - Ir$. Thus if a battery of e.m.f. 12 volts and internal resistance 0.01 ohms delivers a current of 100 A, the terminal p.d.,

$V = 12 - (100)(0.01) = 12 - 1 = 11$ volts.

(iii) When different values of potential difference V, across a cell or power supply are measured for different values of current I, a graph may be plotted as shown in *Fig 6*. Since the e.m.f. E of the cell or power supply is the p.d. across its terminals on no load (i.e. when $I = 0$), then E is as shown by the broken line. Since $V = E - Ir$ then the internal resistance may be calculated from

$$r = \frac{E - V}{I}$$

Fig 6

23 (a) When a current is flowing in the direction shown in *Fig 5* the cell is said to be discharging $(E > V)$.

(b) When a current flows in the opposite direction to that shown in *Fig 5* the cell is said to be charging $(V > E)$.

4

24 A **battery** is a combination of more than one cell. The cells in a battery may be connected in series or in parallel.

(i) For cells connected in series:

   Total e.m.f. = sum of cells' e.m.f.s

   Total internal resistance = sum of cells' internal resistances

(ii) For cells connected in parallel:

If each cell has the same e.m.f. and internal resistance:

   Total e.m.f. = e.m.f. of one cell

   Total internal resistance of n cells = $\frac{1}{n}$ × internal resistance of one cell

B. WORKED PROBLEMS ON SIMPLE D.C. CIRCUITS

Problem 1 What current must flow if 0.24 coulombs is to be transferred in 15 ms?

Since the quantity of electricity, $Q = It$, then

$$I = \frac{Q}{t} = \frac{0.25}{15 \times 10^{-3}} = \frac{0.24 \times 10^3}{15} = \frac{240}{15} = \textbf{16 A}$$

Problem 2 If a current of 10 A flows for four minutes, find the quantity of electricity transferred.

Quantity of electricity, $Q = It$ coulombs
$I = 10$ A; $t = 4 \times 60 = 240$ s
Hence $Q = 10 \times 240 = \textbf{2400 C}$

Problem 3 Find the conductance of a conductor of resistance (a) 10 Ω; (b) 5 kΩ; (c) 100 mΩ.

(a) Conductance $G = \frac{1}{R} = \frac{1}{10}$ siemen = **0.1 S**

(b) $G = \frac{1}{R} = \frac{1}{5 \times 10^3}$ S $= 0.2 \times 10^{-3}$ S = **0.2 mS**

(c) $G = \frac{1}{R} = \frac{1}{100 \times 10^{-3}}$ S $= \frac{10^3}{100}$ S = **10 S**

Problem 4 A source of e.m.f. 15 V supplies a current of 2 A for six minutes. How much energy is provided in this time?

Energy = power × time, and power = voltage × current
Hence energy $= VIt = 15 \times 2 \times (6 \times 60) = 10\ 800$ W s or J
$$= \textbf{10.8 kJ}$$

Problem 5 Electrical equipment in an office takes a current of 13 A from a 240 V supply. Estimate the cost per week of electricity if the equipment is used for 30 hours each week and 1 kW h of energy costs 5p.

Power = VI watts = 240 × 13 = 3120 W = 3.12 kW.

Energy used per week = power × time = (3.12 kW) × (30 h)

$$= 93.6 \text{ kW h}$$

Cost at 5p per kW h = 93.6 × 5   = 468p

Hence weekly cost of electricity = **£4.68**

Problem 6 An electric heater consumes 3.6 MJ when connected to a 250 V supply for 40 minutes. Find the power rating of the heater and the current taken from the supply.

$$\text{Power} = \frac{\text{energy}}{\text{time}} = \frac{3.6 \times 10^6}{40 \times 60} \; \frac{\text{J}}{\text{s}} \text{ (or W)} = 1500 \text{ W}$$

i.e. Power rating of heater = **1.5 kW**

Power $P = VI$, thus $I = \dfrac{P}{V} = \dfrac{1500}{250} = 6$ A

Hence the current taken from the supply is **6 A**

Problem 7 A coil has a current of 50 mA flowing through it when the applied voltage is 12 V. What is the resistance of the coil?

Resistance, $R = \dfrac{V}{I} = \dfrac{12}{50 \times 10^{-3}} = \dfrac{12 \times 10^3}{50} = \dfrac{12\,000}{50} = \textbf{240}\ \boldsymbol{\Omega}$

Problem 8 An electric kettle has a resistance of 30 Ω. What current will flow when it is connected to a 240 V supply? Find also the power rating of the kettle.

Current, $I = \dfrac{V}{R} = \dfrac{240}{30} = \textbf{8 A}$

Power, $P = VI = 240 \times 8 = 1920$ W = **1.92 kW** = power rating of kettle.

Problem 9 Find the equivalent resistance for the circuit shown in *Fig 7*.

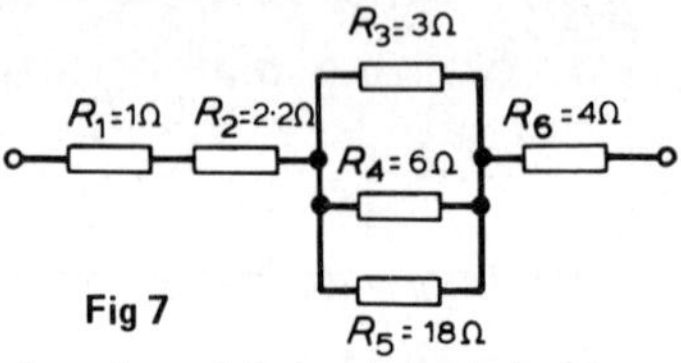

R_3, R_4 and R_5 are connected in parallel and their equivalent resistance R is given by:

$$\frac{1}{R} = \frac{1}{3} + \frac{1}{6} + \frac{1}{18} = \frac{6+3+1}{18} = \frac{10}{18}$$

Hence $R = \dfrac{18}{10} = 1.8\ \Omega$

Fig 7

The circuit is now equivalent to four resistors in series and the equivalent circuit resistance = 1 + 2.2 + 1.8 + 4 = **9 Ω**

Problem 10 Calculate the equivalent resistance between the points A and B for the circuit shown in *Fig 8*.

Combining the two 3 Ω resistors in series, the three 10 Ω resistors in series and the 2.5 Ω, 1 Ω and 1.5 Ω resistors in series gives the simplified equivalent circuit of *Fig 9*. The equivalent resistance R of 6 Ω, 15 Ω and 30 Ω in parallel is given by:

$$\frac{1}{R} = \frac{1}{6} + \frac{1}{15} + \frac{1}{30} = \frac{5+2+1}{30} = \frac{8}{30}$$

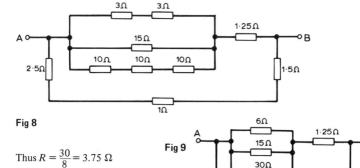

Fig 8

Thus $R = \dfrac{30}{8} = 3.75\ \Omega$

The equivalent circuit is now as shown in
Fig 10. Combining the 3.75 Ω and
1.25 Ω in series gives an equivalent resistance
of 5 Ω. The equivalent resistance R_T of 5 Ω
in parallel with another 5 Ω resistor is given by:

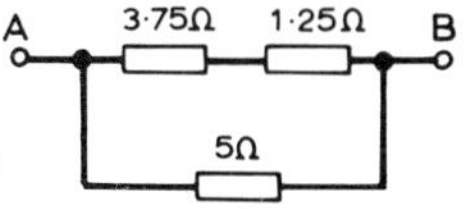

$$R_T = \frac{5 \times 5}{5 + 5} = \frac{25}{10} = 2.5\ \Omega$$

Fig 9

Fig 10

(Note that when two resistors having the same value are connected in parallel the
equivalent resistance will always be half the value of one of the resistors.) **The circuit
of *Fig 8* can thus be replaced by a 2.5 Ω resistor placed between points A and B.**

Problem 11 Determine the equivalent resistance for the series-parallel arrangement shown in *Fig 11*, correct to
two decimal places.

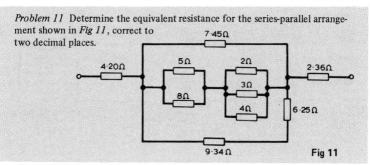

Fig 11

The equivalent resistance of 5 Ω in parallel with 8 Ω is :

$$\frac{5 \times 8}{5 + 8} = \frac{40}{13}\ , \text{ i.e. } 3.077\ \Omega.$$

The equivalent resistance R of 2 Ω, 3 Ω and 4 Ω in parallel is given by:

$$\frac{1}{R} = \frac{1}{2} + \frac{1}{3} + \frac{1}{4} = \frac{6 + 4 + 3}{12} = \frac{13}{12}$$

Hence $R = \dfrac{12}{13} = 0.923\ \Omega$

The equivalent resistance of 9.34 Ω and 6.25 Ω in series is $9.34 + 6.25 = 15.59\ \Omega$.
Thus a simplified circuit diagram is shown in *Fig 12*. It will be seen that 3.077 Ω in
series with 0.923 Ω gives an equivalent resistance of 4.00 Ω. The equivalent resistance
R_x of 7.45 Ω, 4.00 Ω and 15.59 Ω in parallel is given by:

$$\frac{1}{R_x} = \frac{1}{7.45} + \frac{1}{4.00} + \frac{1}{15.59}$$

i.e.

Conductance G_x

$$= 0.134 + 0.250 + 0.064$$
$$= 0.448 \text{ siemens}$$

Since $G_x = \frac{1}{R_x}$, then

$$R_x = \frac{1}{G_x} = \frac{1}{0.448} = 2.23 \ \Omega$$

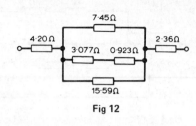

Fig 12

The circuit is now equivalent to three resistors of 4.20 Ω, 2.23 Ω and 2.36 Ω connected in series, which gives an equivalent resistance of 4.20+2.23+2.36 = **8.79 Ω**

Problem 12 Resistances of 10 Ω, 20 Ω and 30 Ω are connected (a) in series and (b) in parallel to a 240 V supply. Calculate the supply current in each case.

(a) The series circuit is shown in
Fig 13. The equivalent resistance
$R_T = 10 \ \Omega + 20 \ \Omega + 30 \ \Omega = 60 \ \Omega$

Supply current $I = \frac{V}{R_T} = \frac{240}{60} = 4$ A

(b) The parallel circuit is shown in *Fig 14*.
The equivalent resistance R_T of 10 Ω, 20 Ω
and 30 Ω resistances connected in parallel is
given by:

$$\frac{1}{R_T} = \frac{1}{10} + \frac{1}{20} + \frac{1}{30} = \frac{6+3+2}{60} = \frac{11}{60}$$

Hence $R_T = \frac{60}{11} \ \Omega$

Supply current $I = \frac{V}{R_T} = \frac{240}{60/11} = \frac{240 \times 11}{60}$

$$= 44 \text{ A}$$

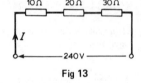

Fig 13

Fig 14

(*Check:* $I_1 = \frac{V}{R_1} = \frac{240}{10} = 24$ A; $I_2 = \frac{V}{R_2} = \frac{240}{20} = 12$ A

$I_3 = \frac{V}{R_3} = \frac{240}{30} = 8$ A

For a parallel circuit $I = I_1 + I_2 + I_3 = 24 + 12 + 8 = $ **44 A**, as above.)

Problem 13 For the series-parallel arrangement shown in *Fig 15* find (a) the supply current, (b) the current flowing through each resistor and (c) the p.d. across each resistor.

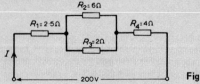

Fig 15

8

(a) The equivalent resistance R_x of R_2 and R_3 in parallel is:

$$R_x = \frac{6 \times 2}{6 + 2} = \frac{12}{8} = 1.5 \ \Omega$$

The equivalent resistance R_T of R_1, R_x and R_4 in series is:

$$R_T = 2.5 + 1.5 + 4 = 8 \ \Omega$$

Supply current $I = \dfrac{V}{R_T} = \dfrac{200}{8} = 25$ A.

(b) The current flowing through R_1 and R_4 is 25 A.

The current flowing through $R_2 = \left(\dfrac{R_3}{R_2 + R_3}\right) I = \left(\dfrac{2}{6 + 2}\right) 25 = \textbf{6.25 A}$

The current flowing through $R_3 = \left(\dfrac{R_2}{R_2 + R_3}\right) I = \left(\dfrac{6}{6 + 2}\right) 25 = \textbf{18.75 A}$

(Note that the currents flowing through R_2 and R_3 must add up to the total current flowing into the parallel arrangement, i.e. 25 A.)

(c) The equivalent circuit of *Fig 15* is shown in *Fig 16*.

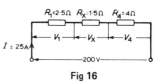

p.d. across R_1, i.e., $V_1 = IR_1 = (25)(2.5)$
         $= 62.5$ V

p.d. across R_x, i.e., $V_x = IR_x = (25)(1.5)$
         $= 37.5$ V

p.d. across R_4, i.e., $V_4 = IR_4 = (25)(4)$
         $= 100$ V

Fig 16

Hence the p.d. across $R_2 = $ p.d. across $R_3 = 37.5$ V

Problem 14 For the circuit shown in *Fig 17* calculate (a) the value of resistor R_x such that the total power dissipated in the circuit is 2.5 kW; (b) the current flowing in each of the four resistors.

Fig 17

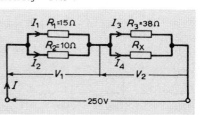

(a) Power dissipated $P = VI$ watts

Hence $2500 = (250)(I)$

$$I = \frac{2500}{250} = 10 \text{ A}$$

From Ohm's law, $R_T = \dfrac{V}{I} = \dfrac{250}{10} = 25 \ \Omega$, where R_T is the equivalent circuit resistance.

The equivalent resistance of R_1 and R_2 in parallel is $\dfrac{15 \times 10}{15 + 10} = \dfrac{150}{25} = 6 \ \Omega$

The equivalent resistance of resistors R_3 and R_x in parallel is equal to $25 \ \Omega - 6 \ \Omega$, i.e., $19 \ \Omega$.

There are three methods whereby R_x can be determined.

Method 1

The voltage $V_1 = IR$, where R is 6 Ω, from above.

9

i.e. $V_1 = (10)(6) = 60$ V

Hence $V_2 = 250$ V $- 60$ V $= 190$ V $=$ p.d. across $R_3 =$ p.d. across R_x.

$I_3 = \dfrac{V_2}{R_3} = \dfrac{190}{38} = 5$ A. Thus $I_4 = 5$ A also, since $I = 10$ A.

Thus $R_x = \dfrac{V_2}{I_4} = \dfrac{190}{5} = 38\ \Omega$

Method 2

Since the equivalent resistance of R_3 and R_x in parallel is $19\ \Omega$,

$19 = \dfrac{38\,R_x}{38 + R_x}$ (i.e. $\dfrac{\text{product}}{\text{sum}}$)

Hence $19\,(38 + R_x) = 38R_x$

$\qquad 722 + 19R_x = 38R_x$

$\qquad\qquad 722 = 38R_x - 19R_x = 19R_x$

Thus $R_x = \dfrac{722}{19} = 38\ \Omega$

Method 3

When two resistors having the same value are connected in parallel the equivalent resistance is always half the value of one of the resistors. Thus, in this case, since $R_T = 19\ \Omega$ and $R_3 = 38\ \Omega$, then $R_x = 38\ \Omega$ could have been deduced on sight.

(b) Current $I_1 = \left(\dfrac{R_2}{R_1 + R_2}\right)I = \left(\dfrac{10}{15 + 10}\right)10 = \left(\dfrac{2}{5}\right)10 = 4$ A

$\qquad$ Current $I_2 = \left(\dfrac{R_1}{R_1 + R_2}\right)I = \left(\dfrac{15}{15 + 10}\right)10 = \left(\dfrac{3}{5}\right)10 = 6$ A

$\qquad$ From part (a), method 1, $I_3 = I_4 = 5$ A

Problem 15 For the arrangement shown in *Fig 18*, find the current I_x.

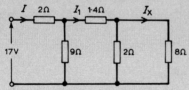

Fig 18

Commencing at the right-hand side of the arrangement shown in *Fig 18*, the circuit is gradually reduced in stages as shown in *Fig 19(a)–(d)*.

From *Fig 19(d)*.

$I = \dfrac{17}{4.25} = 4$ A

From *Fig 19(b)*.

$I_1 = \left(\dfrac{9}{9 + 3}\right)I = \left(\dfrac{9}{12}\right)4 = 3$ A

From *Fig 18*.

$I_x = \left(\dfrac{2}{2 + 8}\right)I_1 = \left(\dfrac{2}{10}\right)3 = 0.6$ A

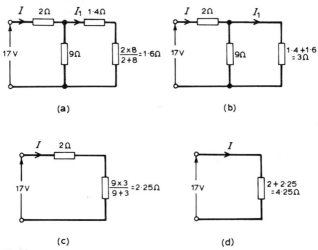

(a)   **(b)**

(c)   **(d)**

Fig 19

Problem 16 A cell has an internal resistance of 0.03 Ω and an e.m.f. of 2.20 V. Calculate its terminal p.d. if it delivers (a) 1 A, (b) 10 A, (c) 40 A.

(a) For 1 A, terminal p.d. $V = E - Ir = 2.20 - (1)(0.03) = $ **2.17 V**
(b) For 10 A, terminal p.d. $V = E - Ir = 2.20 - (10)(0.03) = $ **1.90 V**
(c) For 40 A, terminal p.d. $V = E - Ir = 2.20 - (40)(0.03) = $ **1.00 V**

Problem 17 The voltage at the terminals of a battery is 75 V when no load is connected and 72 V when a load of 60 A is connected. Find the internal resistance of the battery. What would be the terminal voltage when a load taking 40 A is connected?

When no load is connected $E = V$, hence the e.m.f. E of the battery is 75 V.
When a load is connected, the terminal voltage $V = E - Ir$. Hence

$72 = 75 - (60)(r)$ and $60r = 75 - 72 = 3$
$r = \dfrac{3}{60} = \dfrac{1}{20} = $ **0.05 Ω** = internal resistance of the battery.
When a current of 40 A is flowing then

$V = 75 - (40)(0.05) = 75 - 2 = $ **73 V**

Problem 18 A battery consists of 10 cells connected in series, each cell having an e.m.f. of 2 V and an internal resistance of 0.05 Ω. The battery supplies a load R taking 4 A. Find the voltage at the battery terminals and the value of the load R.

For cells connected in series, total e.m.f. = sum of individual e.m.f.s = 20 V.
Total internal resistance = sum of individual internal resistances = 0.5 Ω.
The circuit diagram is shown in *Fig 20*.

11

Voltage at battery terminals

$V = E - Ir = 20 - (4)(0.5)$ or $V = 18$ V

Resistance of load

$R = \dfrac{V}{I} = \dfrac{18}{4} = 4.5 \ \Omega$

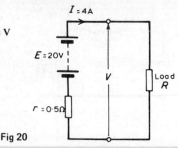

Fig 20

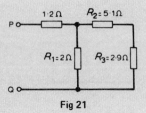

Problem 19 Determine the equivalent resistance of the network shown in *Fig 21*. Hence determine the current taken from the supply when a battery of e.m.f. 12 V and internal resistance 0.2 Ω is connected across the terminals PQ. Find also the current flowing through the 2.9 Ω resistor and the p.d. across the 5.1 Ω resistor.

Fig 21

R_2 in series with R_3 is equivalent to 5.1 Ω + 2.9 Ω, i.e. 8 Ω

R_1 in parallel with 8 Ω gives an equivalent resistance of $\dfrac{2 \times 8}{2 + 8} = 1.6 \ \Omega$

1.6 Ω in series with 1.2 Ω gives an equivalent resistance of 2.8 Ω.
Hence the equivalent resistance of the network shown in *Figure 21* is 2.8 Ω
Fig 22 shows the equivalent resistance connected to the battery.

Current $I = \dfrac{E}{R_T}$, where R_T is the total

circuit resistance (i.e. including the internal resistance r of the battery).

Hence $I = \dfrac{12}{2.8 + 0.2} = \dfrac{12}{3.0} = 4$ A

(Note that in *Fig 22* the resistance 2.8 Ω and 0.2 Ω are connected in series with each other and not in parallel.)

From *Fig 23*, the current flowing through the 2.9 Ω resistor, I_1, is given by

$I_1 = \left(\dfrac{2}{2 + 5.1 + 2.9} \right) (4) = 0.8$ A

The p.d. across the 5.1 Ω resistor is

$V = I_1(5.1) = (0.8)(5.1) = 4.08$ V

Fig 22

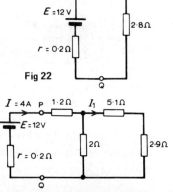

Fig 23

C. FURTHER PROBLEMS ON SIMPLE D.C. CIRCUITS

(a) SHORT ANSWER PROBLEMS

1   Define electric current.

2 Name the unit used to measure (a) the quantity of electricity; (b) resistance; (c) conductance.

3 Define the coulomb.

4 Define electrical energy and name its unit.

5 Define electrical power and name its unit.

6 What do you understand by the term 'potential difference'?

7 What is electromotive force?

8 Write down three formulae for calculating the power in a d.c. circuit.

9 Write down the symbols for the quantities: (a) electrical charge; (b) work; (c) e.m.f.; (d) p.d.

10 State to which units the following abbreviations refer:
(a) A; (b) C; (c) J; (d) N; (e) m.

11 Explain the meaning of these prefixes:
(a) m; (b) k; (c) p; (d) M; (e) μ; (f) n.

12 State Ohm's law.

13 Name three characteristics of a series circuit.

14 Name three characteristics of a parallel circuit.

15 Explain the potential divider circuit.

16 What is a potentiometer?

17 Define internal resistance and terminal p.d. as applied to a voltage source.

18 If a cell has an e.m.f. of E volts, and a terminal p.d. of V volts, when supplying a current of I amperes, the internal resistance r of the cell is given by: $r = \ldots\ldots$

(b) MULTI-CHOICE PROBLEMS (answers on page 172)

1 Which of the following formulae for electrical power is incorrect?
(a) VI; (b) $\frac{V}{I}$; (c) I^2R; (d) $\frac{V^2}{R}$.

2 A resistance of 50 kΩ has a conductance of
(a) 20 S; (b) 0.02 S; (c) 0.02 mS; (d) 20 kS.

3 State which of the following is incorrect:
(a) 1 N = 1 kg m/s^2; (b) 1 V = 1 J/As; (c) 30 mA = 0.03 A; (d) 1 J = 1 N/m.

4 The power dissipated by a resistor of 4 Ω when a current of 5 A passes through it is:
(a) 6.25 W; (b) 20 W; (c) 80 W; (d) 100 W.

5 60 μs is equivalent to: (a) 0.06 s; (b) 0.000 06 s; (c) 1000 minutes; (d) 0.6 s.

6 A current of 3 A flows for 50 h through a 6 Ω resistor. The energy consumed by the resistor is:
(a) 0.9 kW h; (b) 2.7 kW h; (c) 9 kW h; (d) 27 kW h.

7 What must be known in order to calculate the energy used by an electrical appliance?
(a) voltage and current; (b) current and time of operation; (c) power and time of operation; (d) current and quantity of electricity used.

8   If two 4 Ω resistors are placed in series the effective resistance of the circuit is:
    (a) 8 Ω; (b) 4 Ω; (c) 2 Ω; (d) 1 Ω.

9   If two 4 Ω resistors are placed in parallel the effective resistance of the circuit is:
    (a) 8 Ω; (b) 4 Ω; (c) 2 Ω; (d) 1 Ω.

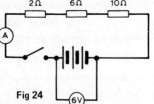

10 With the switch in *Fig 24* closed the
    ammeter reading will indicate:
    (a) 108 A; (b) $\frac{1}{3}$ A; (c) 3 A; (d) $4\frac{3}{5}$ A.

11 A 6 Ω resistor is connected in parallel with the
    three resistors of *Fig 24*. With the switch
    closed the ammeter reading will indicate:
    (a) $\frac{3}{4}$ A; (b) 4 A; (c) $\frac{1}{4}$ A; (d) $1\frac{1}{3}$ A.

Fig 24

12 A 10 Ω resistor is connected in parallel with a 15 Ω resistor and the combination
    is connected in series with a 12 Ω resistor. The equivalent resistance of the circuit
    is: (a) 37 Ω; (b) 18 Ω; (c) 27 Ω; (d) 4 Ω.

13 The terminal voltage of a cell of e.m.f. 2 V and internal resistance 0.1 Ω when
    supplying a current of 5 A will be: (a) 1.5 V; (b) 2 V; (c) 1.9 V; (d) 2.5 V.

14 The effect of connecting an additional parallel load to an electrical supply source
    is to increase the: (a) resistance of the load; (b) voltage of the source; (c) current
    taken from the source; (d) p.d. across the load.

15 The equivalent resistance when a resistor of $\frac{1}{4}$ Ω is connected in parallel with a
    $\frac{1}{5}$ Ω resistor is: (a) $\frac{1}{9}$ Ω; (b) 9 Ω.

(c) CONVENTIONAL PROBLEMS

1   In what time would a current of 10 A transfer a charge of 50 C?   [5 s]

2   A current of 6 A flows for 10 minutes. What charge is transferred?   [3600 C]

3   How long must a current of 100 mA flow so as to transfer a charge of 50 C?
                                                                           [8 min 20 s]

4   Find the conductance of a resistor of resistance (a) 10 Ω; (b) 2 kΩ; (c) 2 mΩ.
                                                       [(a) 0.1 S; (b) 0.5 mS; (c) 500 S]

5   A conductor has a conductance of 50 μS. What is its resistance?   [20 kΩ]

6   An e.m.f. of 250 V is connected across a resistance and the current flowing through
    the resistance is 4 A. What is the power developed?   [1 kW]

7   85.5 J of energy are converted into heat in nine seconds. What power is dissipated?
                                                                           [9.5 W]

8   A current of 4 A flows through a conductor and 10 W is dissipated. What p.d. exists
    across the ends of the conductor?   [2.5 V]

9   Find the power dissipated when:
    (a) a current of 5 mA flows through a resistance of 20 kΩ;
    (b) a voltage of 400 V is applied across a 120 kΩ resistor;
    (c) a voltage applied to a resistor is 10 kV and the current flow is 4 mA.
                                           [(a) 0.5 W; (b) $1\frac{1}{3}$ W; (c) 40 W]

10 A battery of e.m.f. 15 V supplies a current of 2 A for 5 min. How much energy is
    supplied in this time?   [9 kJ]

11 An electric heater takes 7.5 A from a 250 V supply. Find the annual cost, to the nearest pence, if the heater is used an average of 25 hours per week for 48 weeks. Assume that 1 kW h of energy costs 5p. [£112.50]

12 A d.c. electric motor consumes 72 MJ when connected to a 400 V supply for 2 h 30 min. Find the power rating of the motor and the current taken from the supply. [8 kW; 20 A]

13 Determine what voltage must be applied to a 2 kΩ resistor in order that a current of 10 mA may flow. [20 V]

14 The hot resistance of a 240 V filament lamp is 960 Ω. Find the current taken by the lamp and its power rating. [0.25 A; 60 W]

15 Determine the p.d. across a 240 Ω resistance when 12.5 mA is flowing through it. [3 V]

16 Find the resistance of an electric fire which takes a maximum current of 13 A from a 240 V supply. Find also the power rating of the fire. [18.46 Ω; 3.12 kW]

17 What is the resistance of a coil which draws a current of 80 mA from a 120 V supply? [1.5 kΩ]

18 Find the equivalent resistance when the following resistances are connected (a) in series; (b) in parallel.
(i) 3 Ω and 2 Ω; (ii) 20 kΩ and 40 kΩ; (iii) 4 Ω, 8 Ω and 16 Ω; (iv) 800 Ω, 4 kΩ and 1500 Ω.

$$\begin{bmatrix} \text{(a) (i) 5 }\Omega\text{; (ii) 60 k}\Omega\text{; (iii) 28 }\Omega\text{; (iv) 6.3 k}\Omega. \\ \text{(b) (i) 1.2 }\Omega\text{; (ii) 13}\tfrac{1}{3}\text{ k}\Omega\text{; (iii) 2}\tfrac{2}{7}\text{ }\Omega\text{; (iv) 461.5 }\Omega \end{bmatrix}$$

19 If four similar lamps are connected in parallel and the total resistance of the circuit is 150 Ω, find the resistance of one lamp. [600 Ω]

20 An electric circuit has resistances of 2.41 Ω, 3.57 Ω and 5.82 Ω connected in parallel. Find (a) the total circuit conductance; (b) the total circuit resistance.
[(a) 0.867 S; (b) 1.154 Ω]

21 Find the total resistance between terminals A and B of the circuit shown in *Fig 25(a)*. [8 Ω]

22 Find the equivalent resistance between terminals C and D of the circuit shown in *Fig 25(b)*. [27.5 Ω]

23 Determine the equivalent resistance between terminals E and F of the circuit shown in *Fig 25(c)*. [2 Ω]

24 Find the equivalent resistance between terminals G and H of the circuit shown in *Fig 25(d)*. [13.62 Ω]

25 State how four 1 Ω resistors must be connected to give an overall resistance of:
(a) $\tfrac{1}{4}$ Ω; (b) 1$\tfrac{1}{3}$ Ω; (c) 1 Ω; (d) 2$\tfrac{1}{2}$ Ω.

$$\begin{bmatrix} \text{(a)} & \text{Four in parallel; (b) three in parallel, in series with one;} \\ \text{(c)} & \text{two in parallel, in series with another two in parallel} \\ & \text{(or two in series, in parallel with another two in series);} \\ \text{(d)} & \text{two in parallel, in series with two in series.} \end{bmatrix}$$

26 Resistors of 20 Ω, 20 Ω and 30 Ω are connected in parallel. What resistance must

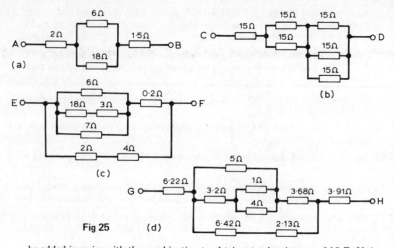

Fig 25

(a) (b) (c) (d)

be added in series with the combination to obtain a total resistance of 10 Ω. If the complete circuit expends a power of 0.36 kW, find the total current flowing.

[2.5 Ω; 6 A]

27 (a) Calculate the current flowing in the 30 Ω resistor shown in *Fig 26*.

(b) What additional value of resistance would have to be placed in parallel with the 20 Ω and 30 Ω resistors to change the supply current to 8 A, the supply voltage remaining constant.

[(a) 1.6 A; (b) 6 Ω]

Fig 26

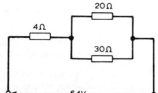

28 For the circuit shown in *Fig 27*, find (a) V_1, (b) V_2, without calculating the current flowing.

[(a) 30 V; (b) 42 V]

29 Determine the currents and voltages indicated in the circuit shown in *Fig 28*.

$$\begin{bmatrix} I_1 = 5 \text{ A} \\ I_2 = 2.5 \text{ A}; \\ I_3 = 1\frac{2}{3} \text{ A}; \\ I_4 = \frac{5}{6} \text{ A}; \\ I_5 = 3 \text{ A}; \\ I_6 = 2 \text{ A}; \\ V_1 = 20 \text{ V}; \\ V_2 = 5 \text{ V}; \\ V_3 = 6 \text{ V} \end{bmatrix}$$

Fig 27

Fig 28

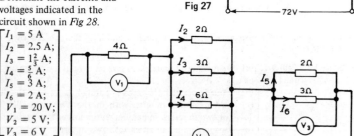

30 *Fig 29* shows part of an electric circuit. Find the value of resistor R and the reading on the ammeter and voltmeters.
$$\begin{bmatrix} R = 18\ \Omega;\ 1.5\ \text{A};\\ V_1 = 15\ \text{V};\\ V_2 = 18\ \text{V} \end{bmatrix}$$

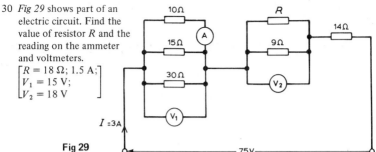

Fig 29

31 A resistor R_x ohms is connected in series with two parallel-connected resistors each of resistance 8 Ω. When the combination is connected across a 280 V supply the power taken by each of the 8 Ω resistors is 392 W. Calculate (a) the resistance of R_x; (b) the single resistance which would take the same power as the series-parallel arrangement.
[(a) 16 Ω; (b) 20 Ω]

32 Find current I in *Fig 30*.
[1.8 A] **Fig 30**

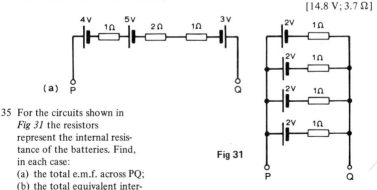

33 A cell has an internal resistance of 0.06 Ω and an e.m.f. of 2.18 V. Find the terminal voltage if it delivers (a) 0.5 A; (b) 1 A; (c) 20 A.
[(a) 2.15 V; (b) 2.12 V; (c) 0.98 V]

34 A battery of e.m.f. 18 V and internal resistance 0.8 Ω supplies a load of 4 A. Find the voltage at the battery terminals and the resistance of the load.
[14.8 V; 3.7 Ω]

35 For the circuits shown in *Fig 31* the resistors represent the internal resistance of the batteries. Find, in each case:
(a) the total e.m.f. across PQ;
(b) the total equivalent internal resistances of the batteries.
[(a) (i) 6 V; (ii) 2 V; (b) (i) 4 Ω; (ii) 0.25 Ω]

Fig 31

(b)

36 The voltage at the terminals of a battery is 52 V when no load is connected and

17

48.8 V when a load taking 80 A is connected. Find the internal resistance of the battery. What would be the terminal voltage when a load taking 20 A is connected?

[0.04 Ω; 51.2 V]

37 A battery of e.m.f. 36.9 V and internal resistance 0.6 Ω is connected to a circuit consisting of a resistance of 1.5 Ω in series with two resistors of 3 Ω and 6 Ω in parallel. Calculate the total current in the circuit, the current flowing through the 6 Ω resistor, the battery terminal p.d. and the volt drop across each resistor.

[9 A; 3 A; 31.5 V; 13.5 V; 18 V; 18 V]

38 A battery consists of four cells connected in series, each having an e.m.f. of 1.28 V and an internal resistance of 0.1 Ω. Across the terminals of the battery are two parallel resistors, $R_1 = 8$ Ω and $R_2 = 24$ Ω. Calculate the current taken by each of the resistors and the energy dissipated in the resistances, in joules, if the current flows for 3½ minutes.

$[I_1 = 0.6; I_2 = 0.2$ A; $W = 806.4$ J]

39 In *Fig 32*, find the total resistance measured between the points A and B. If a battery of e.m.f. 80 V and internal resistance 1 Ω is connected across AB, find the current in each resistor and the p.d. across R_3.

$\begin{bmatrix} 19\ \Omega; I_1 = 4\ \text{A}; I_2 = 0.8\ \text{A}; \\ I_3 = 3.2\ \text{A}; 32\ \text{V} \end{bmatrix}$

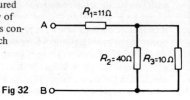

Fig 32

2 Electromagnetism

A. MAIN POINTS CONCERNED WITH ELECTROMAGNETISM

1 A **permanent magnet** is a piece of ferromagnetic material (such as iron, nickel or cobalt) which has properties of attracting other pieces of these materials.
2 The area around a magnet is called the **magnetic field** and it is in this area that the effects of the **magnetic force** produced by the magnet can be detected.
3 The magnetic field of a bar magnet can be represented pictorially by the 'lines of force' (or lines of 'magnetic flux' as they are called) as shown in *Fig 1*. Such a field pattern can be produced by placing iron filings in the vicinity of the magnet.

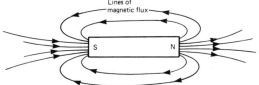

Lines of magnetic flux

Fig 1

The field direction at any point is taken as that in which the north-seeking pole of a compass needle points when suspended in the field. External to the magnet the direction of the field is north to south.

4 The laws of magnetic attraction and repulsion can be demonstrated by using two bar magnets. In *Fig 2(a)*, with **unlike poles** adjacent, **attraction** occurs. In *Fig 2(b)*, with **like poles** adjacent, **repulsion** occurs.

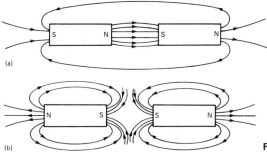

(a)

(b) **Fig 2**

5   Magnetic fields are produced by electric currents as well as by permanent magnets. The field forms a circular pattern with the current-carrying conductor at the centre. The effect is portrayed in *Fig 3* where the convention adopted is:

Current flowing
away from viewer
(a)

Current flowing
towards viewer
(b)

Fig 3

(a) current flowing **away** from the viewer is shown by   – can be thought of as the feathered end of the shaft of an arrow;

(b) current flowing **towards** the viewer is shown by   – can be thought of as the tip of an arrow.

6   The **direction** of the fields in *Fig 3* is remembered by the **screw rule** which states: 'If a normal right-hand thread screw is screwed along the conductor in the direction of the current, the direction of rotation of the screw is in the direction of the magnetic field.'

7   A magnetic field produced by a long coil, or **solenoid**, is shown in *Fig 4* and is seen to be similar to that of a bar magnet shown in *Fig 1*. If the solenoid is wound on an iron bar an even stronger field is produced. The **direction** of the field produced by current *I* is determined by a compass and is remembered by either:

(a) the **screw rule**, which states that if a normal right-hand thread screw is placed along the axis of the solenoid and is screwed in the direction of the current it moves in the direction of the magnetic field inside of the solenoid (i.e. points in the direction of the north pole), or

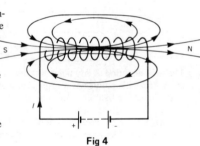

Fig 4

(b) the **grip rule**, which states that if the coil is gripped with the right hand with the fingers pointing in the direction of the current, then the thumb, outstretched parallel to the axis of the solenoid, points in the direction of the magnetic field inside the solenoid (i.e. points in the direction of the north pole).

8   An **electromagnet**, which is a solenoid wound on an iron core, provides the basis of many items of electrical equipment, examples including an electric bell, relays, lifting magnets and telephone receivers.

9   (i) **Magnetic flux** is the amount of magnetic field (or the number of lines of force) produced by a magnetic source.

   (ii) The symbol for magnetic flux is Φ (Greek letter 'phi').

   (iii) The unit of magnetic flux is the **weber, Wb**.

10 (i) Magnetic flux density is the amount of flux passing through a defined area that is perpendicular to the direction of the flux.

Magnetic flux density $= \dfrac{\textbf{magnetic flux}}{\textbf{area}}$

(ii) The symbol for magnetic flux density is B.

(iii) The unit of magnetic flux density is the tesla, T, where $1\ T = 1\ Wb/m^2$.

Hence $$B = \frac{\Phi}{A}\ \text{tesla}$$, where $A\ m^2$ is the area.

(See *Problems 1 and 2*)

11 (i) If a current-carrying conductor is placed in a magnetic field produced by permanent magnets, then the fields due to the current-carrying conductor and the permanent magnets interact and cause a force to be exerted on the conductor. The force on the current-carrying conductor in a magnetic field depends upon:

(a) the flux density of the field, B teslas;

(b) the strength of the current, I amperes;

(c) the length of the conductor perpendicular to the magnetic field, l metre; and

(d) the directions of the field and the current.

(ii) When the magnetic field, the current and the conductor are mutually at right angles then:

Force $F = BIl$ newtons

(iii) When the conductor and the field are at an angle $\theta°$ to each other then:

Force $F = BIl \sin \theta$ newtons

(iv) Since when the magnetic field, current and conductor are mutually at right angles, $F = BIl$, the magnetic flux density B may be defined by $B = \dfrac{F}{Il}$,

i.e. the flux density is 1 T if the force exerted on 1 m of a conductor when the conductor carries a current of 1 A is 1 N

12 If the current-carrying conductor shown in *Fig 3(a)*, is placed in the magnetic field shown in *Fig 5(a)*, then the two fields interact and cause a force to be exerted on the conductor as shown in *Fig 5(b)*. The field is strengthened above the conductor and weakened below, thus tending to move the conductor downwards. This is the basic principle of operation of the electric motor (see *Problem 8*) and the moving-coil instrument (see *Problem 9*)

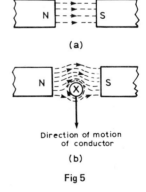

(a)

Direction of motion of conductor

(b)

Fig 5

13 The direction of the force exerted on a conductor can be predetermined by using Fleming's left-hand rule (often called the motor rule), which states:
'Let the thumb, first finger and second finger of the left hand be extended such that they are all at right-angles to each other,' (as shown in *Fig 6*). 'If the first finger points in the direction of the magnetic field, the second finger points in the

21

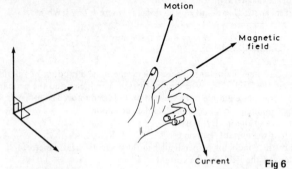

Motion

Magnetic field

Current   Fig 6

direction of the current, then the thumb will point in the direction of the motion of the conductor.'

Summarising:

First finger – Field
SeCond finger – Current
ThuMb - Motion

(See *Problems 3 to 7*)

14 When a charge of Q coulombs is moving at a velocity of v m/s in a magnetic field of flux density B teslas, the charge moving perpendicular to the field, then the magnitude of the force F exerted on the charge is given by:

$$\boxed{F = QvB \text{ newtons}}$$

(See *Problem 10*)

B. WORKED PROBLEMS ON ELECTROMAGNETISM

Problem 1 A magnetic pole face has a rectangular section having dimensions 200 mm by 100 mm. If the total flux emerging from the pole is 150 μWb, calculate the flux density.

Flux $\Phi = 150 \ \mu\text{Wb} = 150 \times 10^{-6}$ Wb
Cross sectional area $A = 200 \times 100 = 20\ 000 \ \text{mm}^2 = 20\ 000 \times 10^{-6} \ \text{m}^2$

Flux density $B = \dfrac{\Phi}{A} = \dfrac{150 \times 10^{-6}}{20\ 000 \times 10^{-6}} = \textbf{0.0075 T or 7.5 mT}$

Problem 2 The maximum working flux density of a lifting electromagnet is 1.8 T and the effective area of a pole face is circular in cross-section. If the total magnetic flux produced is 353 mWb, determine the radius of the pole face.

Flux density $B = 1.8$ T; flux $\Phi = 353$ mWb $= 353 \times 10^{-3}$ Wb

Since $B = \dfrac{\Phi}{A}$, cross-sectional area $A = \dfrac{\Phi}{B} = \dfrac{353 \times 10^{-3}}{1.8} \ \text{m}^2 = 0.1961 \ \text{m}^2$

The pole face is circular, hence area $= \pi r^2$, where r is the radius.

Hence $\pi r^2 = 0.1961$

from which $r^2 = \dfrac{0.1961}{\pi}$ and radius $r = \sqrt{\left(\dfrac{0.1961}{\pi}\right)} = 0.250 \text{ m}^2$

i.e. the radius of the pole face is **250 mm**

Problem 3 A conductor carries a current of 20 A and is at right-angles to a magnetic field having a flux density of 0.9 T. If the length of the conductor in the field is 30 cm, calculate the force acting on the conductor. Determine also the value of the force if the conductor is inclined at an angle of 30° to the direction of the field.

$B = 0.9$ T; $I = 20$ A; $l = 30$ cm $= 0.30$ m.
Force $F = BIl = (0.9)(20)(0.30)$ newtons when the conductor is at right-angles to the field, as shown in *Fig 7(a)*, i.e. $F = 5.4$ N.

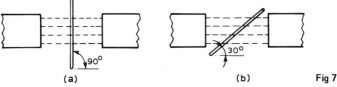

(a)   **(b)**   **Fig 7**

When the conductor is incliend at 30° to the field, as shown in *Fig 7(b)*, then
Force $F = BIl \sin \theta$
i.e. $F = (0.9)(20)(0.30) \sin 30°$
    $F = 2.7$ N

Problem 4 Determine the current required in a 400 mm length of conductor of an electric motor, when the conductor is situated at right-angles to a magnetic field of flux density 1.2 T, if a force of 1.92 N is to be exerted on the conductor. If the conductor is vertical, the current flowing downwards and the direction of the magnetic field is from left to right, what is the direction of the force?

$F = 1.92$ N; $l = 400$ mm $= 0.40$ m; $B = 1.2$ T.

Since $F = BIl$, $\quad I = \dfrac{F}{Bl}$

Hence current $I = \dfrac{1.92}{(1.2)(0.4)} = 4$ A

If the current flows downwards, the direction of its magnetic field due to the current alone will be clockwise when viewed from above. The lines of flux will reinforce (i.e. strengthen) the main magnetic field at the back of the conductor and will be in opposition in the front (i.e. weaken the field). **Hence the force on the conductor will be from back to front (i.e. toward the viewer).** This direction may also have been deduced using Fleming's left-hand rule.

Problem 5 A conductor 350 mm long carries a current of 10 A and is at right-angles to a magnetic field lying between two circular pole faces each of radius 60 mm. If the total flux between the pole faces is 0.5 mWb, calculate the magnitude of the force exerted on the conductor.

$l = 350$ mm $= 0.35$ m; $I = 10$ A; Area of pole face $A = \pi r^2 = \pi(0.06)^2$ m^2;
$\Phi = 0.5$ mWb $= 0.5 \times 10^{-3}$ Wb

Force $F = BIl$, and $B = \dfrac{\Phi}{A}$

Hence $F = \dfrac{\Phi}{A} Il = \dfrac{(0.5 \times 10^{-3})}{\pi(0.06)^2} (10)(0.35)$ newtons

i.e., **force = 0.155 N**

Problem 6 With reference to *Fig 8* determine (a) the direction of the force on the conductor in *Fig 8(a)*; (b) the direction of the force on the conductor in *Fig 8(b)*; (c) the direction of the current in *Fig 8(c)*; (d) the polarity of the magnetic system in *Fig 8(d)*.

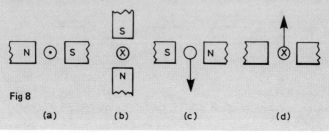

Fig 8

(a) (b) (c) (d)

(a) The direction of the main magnetic field is from north to south, i.e. left to right. The current is flowing towards the viewer, and using the screw rule, the direction of the field is anticlockwise. Hence either by Fleming's left-hand rule, or by sketching the interacting magnetic field as shown in *Figure 9(a)*, the direction of the force on the conductor is seen to be upward.

(b) Using a similar method to part (a) it is seen that the force on the conductor is to the right, see *Fig 9(b)*.

(c) Using Fleming's left-hand rule, or by sketching as in *Fig 9(c)*, it is seen that the current is toward the viewer, i.e. out of the paper.

(d) Similar to part (c), the polarity of the magnetic system is as shown in *Fig 9(d)*.

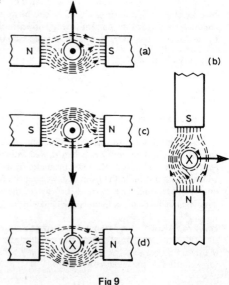

Fig 9

Problem 7 A coil is wound on a rectangular former of width 24 mm and length 30 mm. The former is pivoted about an axis passing through the middle of the two shorter sides and is placed in a uniform magnetic field of flux density 0.8 T, the axis being perpendicular to the field. If the coil carries a current of 50 mA, determine the force on each coil side (a) for a single-turn coil; (b) for a coil wound with 300 turns.

(a) Flux density $B = 0.8$ T; length of conductor lying at right-angles to field $l = 30 \text{ mm} = 30 \times 10^{-3}$ m; current $I = 50 \text{ mA} = 50 \times 10^{-3}$ A.
For a single-turn coil, force on each coil side
$F = BIl = 0.8 \times 50 \times 10^{-3} \times 30 \times 10^{-3} =$ **1.2×10^{-3} N, or 0.0012 N**

(b) When there are 300 turns on the coil there are effectively 300 parallel conductors each carrying a current of 50 mA. Thus the total force produced by the current is 300 times that for a single-turn coil.
Hence force on coil side $F = 300\, BIl = 300 \times 0.0012 =$ **0.36 N**

Problem 8 Describe briefly, with an appropriate sketch, the principle of operation of a simple d.c. motor.

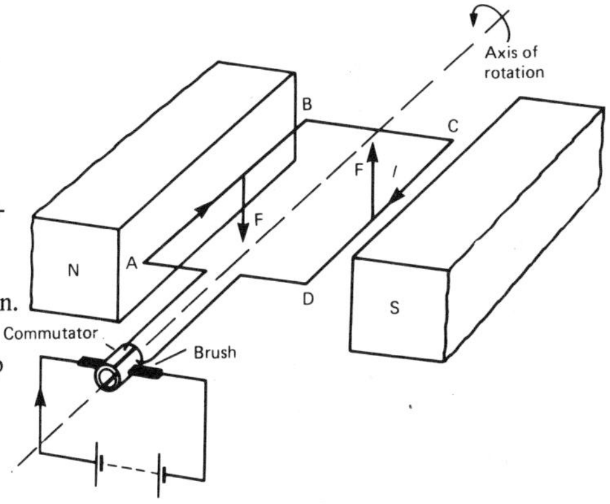

Fig 10

A rectangular coil which is free to rotate about a fixed axis is shown placed inside a magnetic field produced by permanent magnets in *Fig 10.* A direct current is fed into the coil via carbon brushes bearing on a commutator, which consists of a metal ring split into two halves separated by insulation. When current flows in the coil a magnetic field is set up around the coil which interacts with the magnetic field produced by the magnets. This causes a force F to be exerted on the current-carrying conductor which, by Fleming's left-hand rule (see para. 13), is downward between points A and B and upward between C and D for the current direction shown. This causes a torque and the coil rotates anticlockwise. When the coil has turned through 90° from the position shown in *Fig 10* the brushes connected to the positive and negative terminals of the supply make contact with different halves of the commutator ring, thus reversing the direction of the current flow in the conductor. If the current is not reversed and the coil rotates past this position the forces acting on it change direction and it rotates in the opposite direction thus never making more than half a revolution. The current direction is reversed every time the coil swings through the vertical position and thus the coil rotates anticlockwise for as long as the current flows. This is the principle of operation of a d.c. motor which is thus a device that takes in electrical energy and converts it into mechanical energy.

A moving-coil instrument operates on the motor principle. When a conductor carrying current is placed in a magnetic field, a force F is exerted on the conductor, given by $F = BIl$. If the flux density B is made constant (by using permanent magnets) and the conductor is a fixed length (say, a coil) then the force will depend only on the current flowing in the conductor.

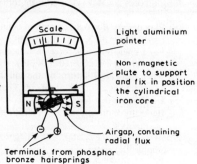

(a)

In a moving-coil instrument a coil is placed centrally in the gap between shaped pole pieces as shown by the front elevation in *Fig 11(a)*. (The airgap is kept as small as possible, although for clarity it is shown exaggerated in *Fig 11*.) The coil is supported by steel pivots, resting in jewel bearings, on a cylindrical iron core. Current is led into and out of the coil by two phosphor bronze spiral hairsprings which are wound in opposite directions to minimise the effect of temperature change and to limit the coil swing (i.e. to **control** the movement) and return the movement to zero position when no current flows. Current flowing in the coil produces forces as shown in *Fig 11(b)*, the directions being obtained by Fleming's left-hand rule. The two forces, F_A and F_B, produce a torque which will move the coil in clockwise direction, i.e. move the pointer from left to right. Since force is proportional to current the scale is linear.

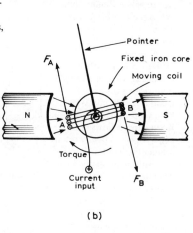

(b)

Fig 11

When the aluminium frame, on which the coil is wound, is rotated between the poles of the magnet, small currents (called eddy currents) are induced into the frame, and this provides automatically the necessary **damping** of the system due to the reluctance of the former to move within the magnetic field.

The moving-coil instrument will measure only direct current or voltage and the terminals are marked positive and negative to ensure that the current passes through the coil in the correct direction to deflect the pointer 'up the scale'.

The range of this sensitive instrument is extended by using shunts and multipliers (see Chapter 5, para. 8).

Problem 10 An electron in a television tube has a charge of 1.6×10^{-19} coulombs and travels at 3×10^7 m/s perpendicular to a field of flux density 18.5 μT. Determine the force exerted on the electron in the field.

From para. 14, force $F = QvB$ newtons, where Q=charge in coulombs=1.6×10^{-19} C; v = velocity of charge = 3×10^7 m/s; and B = flux density = 18.5×10^{-6} T
Hence force on electron $F = 1.6 \times 10^{-19} \times 3 \times 10^7 \times 18.5 \times 10^{-6}$
$= 1.6 \times 3 \times 18.5 \times 10^{-18} = 88.8 \times 10^{-18} = \mathbf{8.88 \times 10^{-17}}$ **N**

C. FURTHER PROBLEMS ON ELECTROMAGNETISM

(a) SHORT ANSWER PROBLEMS

1 What is a permanent magnet?

2 Sketch the pattern of the magnetic field associated with a bar magnet. Mark the direction of the field.

3 The direction of the magnetic field around a current-carrying conductor may be remembered using the . . . . . . rule.

4 Sketch the magnetic field pattern associated with a solenoid connected to a battery and wound on an iron bar. Show the direction of the field.

5 Name three applications of electromagnetism.

6 State what happens when a current-carrying conductor is placed in a magnetic field between two magnets.

7 Define magnetic flux.

8 The symbol for magnetic flux is . . . . . . and the unit of flux is the . . . .

9 Define magnetic flux density.

10 The symbol for magnetic flux density is . . . . . . and the unit of flux density is . . . .

11 The force on a current-carrying conductor in a magnetic field depends on four factors. Name them.

12 The direction of the force on a conductor in a magnetic field may be predetermined using Fleming's . . . . . . rule.

(b) MULTI-CHOICE PROBLEMS (answers on page 172)

1 The unit of magnetic flux density is the:
(a) weber; (b) weber per metre; (c) ampere per metre; (d) tesla.

2 The total flux in the core of an electrical machine is 20 mWb and its flux density is 1 T. The cross-sectional area of the core is:
(a) 0.05 m^2; (b) 0.02 m^2; (c) 20 m^2; (d) 50 m^2.

3 A conductor carries a current of 10 A at right-angles to a magnetic field having a flux density of 500 mT. If the length of the conductor in the field is 20 cm, the force on the conductor is:
(a) 100 kN; (b) 1 kN; (c) 100 N; (d) 1 N.

4 If a conductor is horizontal, the current flowing from left to right and the direction of the surrounding magnetic field is from above to below, the force exerted on the conductor is:
(a) from left to right; (b) from below to above; (c) away from the viewer; (d) towards the viewer.

5 For the current-carrying conductor lying in the magnetic field shown in *Fig 12(a)*, the direction of the force on the conductor is (a) to the left; (b) upwards; (c) to the right; (d) downward.

6 For the current-carrying conductor lying in the magnetic field shown in *Fig 12(b)* the direction of the current in the conductor is (a) toward the viewer; (b) away from the viewer.

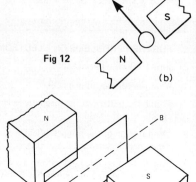

Fig 12

7 *Fig 13* shows a rectangular coil of wire placed in a magnetic field and free to rotate about axis AB. If current flows into the coil at C, the coil will:
(a) commence to rotate anticlockwise;
(b) commence to rotate clockwise;
(c) remain in the vertical position;
(d) experience a force towards the north pole.

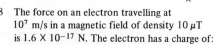

Fig 13

8 The force on an electron travelling at 10^7 m/s in a magnetic field of density 10 μT is 1.6×10^{-17} N. The electron has a charge of:
(a) 1.6×10^{-28} C; (b) 1.6×10^{-15} C; (c) 1.6×10^{-19} C; (d) 1.6×10^{-25} C.

(c) CONVENTIONAL PROBLEMS

1 Determine the flux density in a magnetic field of cross-sectional area 20 cm² having a flux of 3 mWb. [1.5 T]

2 Find the total flux emerging from a magnetic pole face having dimensions 50 mm by 60 mm, if the flux density is 0.9 T. [2.7 mWb]

3 An electromagnet of square cross-section produces a flux density of 0.45 T. If the magnetic flux is 720 μWb find the dimensions of the electromagnet cross-section. [40 mm by 40 mm]

4 The pole core of an electrical machine has a circular cross-section of diameter 120 mm. Determine the flux density if the flux in the core is 7.5 mWb. [0.663 T]

5 A conductor carries a current of 70 A at right-angles to a magnetic field having a flux density of 1.5 T. If the length of the conductor in the field is 200 mm calculate the force acting on the conductor. What is the force when the conductor and field are at an angle of 45°? [21.0 N; 14.8 N]

6 Calculate the current required in a 240 mm length of conductor of a d.c. motor

28

when the conductor is situated at right-angles to the magnetic field of flux density 1.25 T, if a force of 1.20 N is to be exerted on the conductor. [4.0 A]

7 A conductor 30 cm long is situated at right-angles to a magnetic field. Calculate the strength of the magnetic field if a current of 15 A in the conductor produces a force on it of 3.6 N. [0.80 T]

8 A conductor 300 mm long carries a current of 13 A and is at right-angles to a magnetic field between two circular pole faces, each of diameter 80 mm. If the total flux between the pole faces is 0.75 mWb calculate the force exerted on the conductor. [0.582 N]

9 (a) A 400 mm length of conductor carrying a current of 25 A is situated at right-angles to a magnetic field between two poles of an electric motor. The poles have a circular cross-section. If the force exerted on the conductor is 80 N and the total flux between the pole faces is 1.27 mWb, determine the diameter of a pole face.
(b) If the conductor in part (a) is vertical, the current flowing downwards and the direction of the magnetic field is from left to right, what is the direction of the 80 N force? [(a) 14.2 mm; (b) toward the viewer]

10 A coil is wound uniformly on a former having a width of 18 mm and a length of 25 mm. The former is pivoted about an axis passing through the middle of the two shorter sides and is placed in a uniform magnetic field of flux density 0.75 T, the axis being perpendicular to the field. If the coil carries a current of 120 mA, determine the force exerted on each coil side, (a) for a single-turn coil; (b) for a coil wound with 400 turns. [(a) 2.25×10^{-3} N; (b) 0.9 N]

11 *Fig 14* shows a simplified diagram of a section through the coil of a moving-coil instrument. For the direction of current flux shown in the coil determine the direction that the pointer will move.
[To the right]

Fig 14

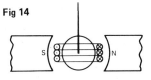

12 Explain, with the aid of a sketch, the action of a simplified d.c. motor.

13 Sketch and label the movement of a moving-coil instrument. Briefly explain the principle of operation of such an instrument.

14 Calculate the force exerted on a charge of 2×10^{-18} C travelling at 2×10^6 m/s perpendicular to a field of density 2×10^{-7} T. [8×10^{-19} N]

15 Determine the speed of a 10^{-19} C charge travelling perpendicular to a field of flux density 10^{-7} T, if the force on the charge is 10^{-20} N. [10^6 m/s]

3 Electromagnetic induction

A. MAIN POINTS CONCERNED WITH ELECTROMAGNETIC INDUCTION

1 When a conductor is moved across a magnetic field, an electromotive force (e.m.f.) is produced in the conductor. If the conductor forms part of a closed circuit then the e.m.f. produced causes an electric current to flow round the circuit. Hence an e.m.f. (and thus current) is 'induced' in the conductor as a result of its movement across the magnetic field. This effect is known as **'electromagnetic induction'** (see *Problem 1*).

2 **Faraday's laws of electromagnetic induction state:**
  (i) 'An induced e.m.f. is set up whenever the magnetic field linking that circuit changes.'
  (ii) 'The magnitude of the induced e.m.f. in any circuit is proportional to the rate of change of the magnetic flux linking the circuit.'

3 **Lenz's law states:**
  'The direction of an induced e.m.f. is always such as to oppose the effect producing it.

4 An alternative method to Lenz's law of determining relative directions is given by **Fleming's Right-hand rule** (often called the gene**R**ator rule) which states:
  'Let the thumb, first finger and second finger of the right hand be extended such that they are all at right angles to each other' (as shown in *Fig 1*). 'If the first finger points in the direction of the magnetic field, the thumb points in the direction of motion of the conductor relative to the magnetic field, then the second finger will point in the direction of the induced e.m.f.'
  Summarising:

  First finger – **F**ield
  Thu**M**b – **M**otion
  S**E**cond finger – **E**.m.f.

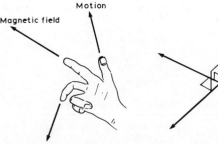

Fig 1

Motion

Magnetic field

Induced E.M.F

30

5   In a generator, conductors forming an electric circuit are made to move through a
    magnetic field. By Faraday's law an e.m.f. is induced in the conductors and thus a
    source of e.m.f. is created. A generator converts mechanical energy into electrical
    energy. (The action of a simple a.c. generator is described in Chapter 4.)
6   The induced e.m.f. E set up between the ends of the conductor shown in
    Fig 2 is given by: $E = Blv$ volts, where B the flux density, is measured
    in teslas, l the length of conductor in the magnetic field, is measured in
    metres, and v the conductor velocity, is measured in metres per second. If

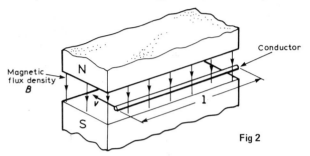

Fig 2

    the conductor moves at an angle $\theta°$ to the magnetic field (instead of at
    90° as assumed above) then

$$E = Blv \sin \theta$$

    (see *Problems 2 to 6*).
7   **Inductance** is the name given to the property of a circuit whereby there is an e.m.f.
    induced into the circuit by the change of flux linkages produced by a current change.
    (i) When the e.m.f. is induced in the same circuit as that in which the current is
    changing, the property is called **self inductance, L.**
    (ii) When the e.m.f. is induced in a circuit by a change of flux due to current chang-
    ing in an adjacent circuit, the property is called **mutual inductance, M.**
8   The unit of inductance is the **henry, H.**
    'A circuit has an inductance of one henry when an e.m.f. of one volt is induced in it
    by a current changing at the rate of one ampere per second.'
9   (i) Induced e.m.f. in a coil of N turns, $E = N\left(\dfrac{\Delta\Phi}{t}\right)$ **volts,** where $\Delta\Phi$ is the change
    in flux in Webers, and t is the time taken for the flux to change in seconds.

    (ii) Induced e.m.f. in a coil of inductance L henrys, $E = L\left(\dfrac{\Delta I}{t}\right)$ **volts,** where ΔI is
    the change in current in amperes and t is the time taken for the current to change in
    seconds
    (See *Problems 7 to 11*).
10 If a current changing from 0 to I amperes, produces a flux change from 0 to Φ Webers,
    then $\Delta I = I$ and $\Delta\Phi = \Phi$. Then, from para. 9, induced e.m.f. $E = \dfrac{N\Phi}{t} = \dfrac{LI}{t}$, from
    which **inductance of coil,** $L = \dfrac{N\Phi}{I}$ **henrys**
    (See *Problems 12 to 14*).
11 Mutually induced e.m.f. in the second coil, $E_2 = M\left(\dfrac{\Delta I_1}{t}\right)$ volts, where M is the
    mutual inductance between two coils in henrys, ΔI_1 is the change in current in the
    first coil in amperes, and t is the time the current takes to change in the first coil
    in seconds
    (See *Problems 15 to 17*).

31

12 (i) A **transformer** is a device which uses the phenomenon of mutual induction to change the values of alternating voltages and currents. In fact, one of the main advantages of a.c. transmission and distribution is the ease with which an alternating voltage can be increased or decreased by transformers.

(ii) Losses in transformers are generally low and thus efficiency is high. Being static they have a long life and are very reliable.

(iii) Transformers range in size from the miniature units used in electronic applications to the large power transformers used in power stations. The principle of operation is the same for each.

13 A transformer is represented in *Fig 3(a)* as consisting of two electrical circuits linked by a common ferromagnetic core. One coil is termed the **primary winding** which is connected to the supply of electricity, and the other the **secondary winding**, which may be connected to a load. A circuit diagram symbol for a transformer is shown in *Fig 3(b)*.

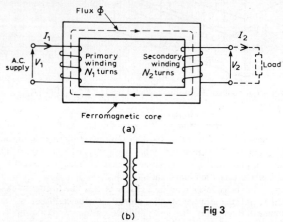

Fig 3

14 **Transformer principle of operation.**

(i) When the secondary is an open-circuit and an alternating voltage V_1 is applied to the primary winding, a small current — called the no-load current I_0 — flows, which sets up a magnetic flux in the core. This alternating flux links with both primary and secondary coils and induces in them e.m.f.s of E_1 and E_2 volts respectively by mutual induction.

(ii) The induced e.m.f. E in a coil of N turns is given by $E = N\left(\dfrac{\Delta\Phi}{t}\right)$ volts, where $\left(\dfrac{\Delta\Phi}{t}\right)$ is the rate of change of flux. In an ideal transformer, the rate of change of flux is the same for both primary and secondary and thus $\dfrac{E_1}{N_1} = \dfrac{E_2}{N_2}$, i.e. **the induced e.m.f. per turn is constant.**

Assuming no losses, $E_1 = V_1$ and $E_2 = V_2$.

Hence $\dfrac{V_1}{N_1} = \dfrac{V_2}{N_2}$ or $\dfrac{V_1}{V_2} = \dfrac{N_1}{N_2}$ (1)

(iii) $\dfrac{V_1}{V_2}$ is called the **voltage ratio** and $\dfrac{N_1}{N_2}$ the **turns ratio**, or the 'transformation

ratio of the transformer. If N_2 is less than N_1 then V_2 is less than V_1 and the device is termed a **step-down transformer**. If N_2 is greater than N_1 then V_2 is greater than V_1 and the device is termed a **step-up transformer**.

(iv) When a load is connected across the secondary winding, a current I_2 flows. In an ideal transformer losses are neglected and a transformer is considered to be 100% efficient. Hence input power = output power, or $V_1 I_1 = V_2 I_2$, i.e. in an ideal transformer, the **primary and secondary ampere-turns are equal**.

Thus $\dfrac{V_1}{V_2} = \dfrac{I_2}{I_1}$ (2)

Combining equations (1) and (2) gives: $\boxed{\dfrac{V_1}{V_2} = \dfrac{N_1}{N_2} = \dfrac{I_2}{I_1}}$ (3)

15 The **rating** of a transformer is stated in terms of the volt-amperes that it can transform without overheating. With reference to *Fig 3(a)*, the transformer rating is either $V_1 I_1$ or $V_2 I_2$, where I_2 is the full-load secondary current. (See *Problems 18 to 22*).

B. WORKED PROBLEMS ON ELECTROMAGNETIC INDUCTION

Problem 1 Briefly describe electromagnetic induction with reference to the movement of a magnet in a coil connected to a meter.

Fig 4(a) shows a coil of wire connected to a centre-zero galvanometer, which is a sensitive ammeter with the zero-current position in the centre of the scale.

(a) When the magnet is moved at constant speed towards the coil (*Fig 4(a)*), a deflection is noted on the galvanometer showing that a current has been produced in the coil.

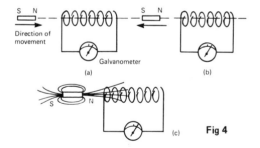

Fig 4

(b) When the magnet is moved at the same speed as in (a) but away from the coil the same deflection is noted but is in the opposite direction (see *Fig 4(b)*).
(c) When the magnet is held stationary even within the coil no deflection is recorded.
(d) When the coil is moved at the same speed as in (a) and the magnet held stationary the same galvanometer deflection is noted.
(e) When the relative speed is, say, doubled, the galvanometer deflection is doubled.
(f) When a stronger magnet is used, a greater galvanometer deflection is noted.
(g) When the number of turns of wire of the coil is increased, a greater galvanometer deflection is noted.

Fig 4(c) shows the magnetic field associated with the magnet. As the magnet is moved towards the coil, the magnetic flux of the magnet moves across, or cuts, the coil. **It is the relative movement of the magnetic flux and the coil that causes an e.m.f. and thus current, to be induced in the coil.** This effect is known as electro-magnetic induction. The laws of electromagnetic induction stated in paras 2 and 3 evolved from experiments such as those described above.

Problem 2 A conductor 300 mm long moves at a uniform speed of 4 m/s at right-angles to a uniform magnetic field of flux density 1.25 T. Determine the current flowing in the conductor when (a) its ends are open-circuited; (b) its ends are connected to a load of 20 Ω resistance.

When a conductor moves in a magnetic field it will have an e.m.f. induced in it but this e.m.f. can only produce a current if there is a closed circuit.

Induced e.m.f. $E = Blv = (1.25) \left(\dfrac{300}{1000} \right) (4) = 1.5$ V

(a) If the ends of the conductor are open circuited **no current will flow** even though 1.5 V has been induced.

(b) From Ohm's law, $I = \dfrac{E}{R} = \dfrac{1.5}{20} = 0.075$ **A or 75 mA**

Problem 3 At what velocity must a conductor 75 mm long cut a magnetic field of flux density 0.6 T if an e.m.f. of 9 V is to be induced in it? Assume the conductor, the field and the direction of motion are mutually perpendicular.

Induced e.m.f. $E = Blv$, hence velocity $v = \dfrac{E}{Bl}$.

Hence $v = \dfrac{9}{(0.6)(75 \times 10^{-3})} = \dfrac{9 \times 10^3}{0.6 \times 75} = $ **200 m/s**

Problem 4 A conductor moves with a velocity of 15 m/s at an angle of (a) 90°, (b) 60°, and (c) 30° to a magnetic field produced between two square-faced poles of side length 2 cm. If the flux leaving a pole face is 5 μWb, find the magnitude of the induced e.m.f. in each case.

$v = 15$ m/s; length of conductor in magnetic field, $l = 2$ cm $= 0.02$ m; $A = 2 \times 2$ cm$^2 = 4 \times 10^{-4}$ m^2; $\Phi = 5 \times 10^{-6}$ Wb.

(a) $E_{90} = Blv \sin 90° = \dfrac{\Phi}{A} lv = \dfrac{(5 \times 10^{-6})}{(4 \times 10^{-4})} (0.02)(15) = $ **3.75 mV**

(b) $E_{60} = Blv \sin 60° = E_{90} \sin 60° = 3.75 \sin 60° = $ **3.25 mV**
(c) $E_{30} = Blv \sin 30° = E_{90} \sin 30° = 3.75 \sin 30° = $ **1.875 mV**

Problem 5 The wing span of a metal aeroplane is 36 m. If the aeroplane is flying at 400 km/h, determine the e.m.f. induced between its wing tips. Assume the vertical component of the earth's magnetic field is 40 μT.

Induced e.m.f. across wing tips, $E = Blv$
$B = 40$ μT $= 40 \times 10^{-6}$ T; $l = 36$ m

$v = 400 \dfrac{\text{km}}{\text{h}} \times 1000 \dfrac{\text{m}}{\text{km}} \times \dfrac{1 \text{ h}}{60 \times 60 \text{ s}} = \dfrac{(400)(1000)}{3600} = \dfrac{4000}{36}$ m/s

Hence $E = (40 \times 10^{-6})(36)\left(\dfrac{4000}{36}\right) = \textbf{0.16 V}$

Problem 6 The diagram shown in *Fig 5* represents the generation of e.m.f.s. Determine (i) the direction in which the conductor has to be moved in *Fig 5(a)*; (ii) the direction of the induced e.m.f. in *Fig 5(b)*; (iii) the polarity of the magnetic system in *Fig 5(c)*.

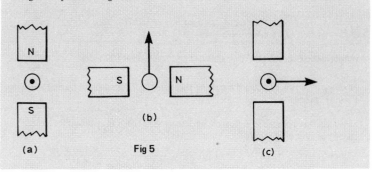

Fig 5

(a)   (b)   (c)

The direction of the e.m.f., and thus the current due to the e.m.f., may be obtained by either Lenz's law or Fleming's Right-hand rule (i.e. GeneRator rule).

(i) Using Lenz's law: The field due to the magnet and the field due to the current-carrying conductor are shown in *Fig 6(a)* and are seen to reinforce to the left of the conductor. Hence the force on the conductor is to the right. However Lenz's law says that the direction of the induced e.m.f. is always such as to oppose the effect producing it. **Thus the conductor will have to be moved to the left.**

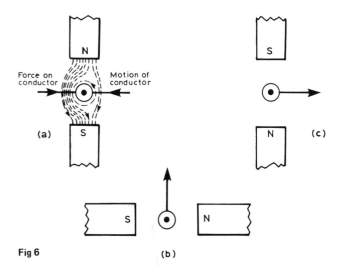

Fig 6   (b)

(ii) Using Fleming's right-hand rule:
First finger – Field, i.e. N → S, or right to left;
ThuMb – Motion, i.e. upwards;
SEcond finger – E.m.f., i.e., **towards the viewer or out of the paper**, as shown in *Figure 6(b)*.
(iii) The polarity of the magnetic system of *Fig 5(c)* is shown in *Fig 6(c)* and is obtained using Fleming's right-hand rule.

Problem 7 Determine the e.m.f. induced in a coil of 200 turns when there is a change of flux of 25 mWb linking with it in 50 ms.

Induced e.m.f. $E = N\left(\dfrac{\Delta\Phi}{t}\right) = (200)\left(\dfrac{25 \times 10^{-3}}{50 \times 10^{-3}}\right) = \mathbf{100\ volts}$

Problem 8 A flux of 400 μWb passing through a 150-turn coil is reversed in 40 ms. Find the average e.m.f. induced.

Since the flux reverses, the flux changes from +400 μWb to −400 μWb, a total change of flux of 800 μWb.

Induced e.m.f. $E = N\left(\dfrac{\Delta\Phi}{t}\right) = (150)\left(\dfrac{800 \times 10^{-6}}{40 \times 10^{-3}}\right) = \dfrac{800 \times 150 \times 10^3}{40 \times 10^6}$

Hence the average e.m.f. induced $E = \mathbf{3\ volts}$

Problem 9 Calculate the e.m.f. induced in a coil of inductance 12 H by a current changing at the rate of 4 A/s.

Induced e.m.f. $E = L\left(\dfrac{\Delta I}{t}\right) = (12)(4) = \mathbf{48\ volts}$

Problem 10 An e.m.f. of 1.5 kV is induced in a coil when a current of 4 A collapses uniformly to zero in 8 ms. Determine the inductance of the coil.

Change in current, $\Delta I = (4-0) = 4$ A; $t = 8$ ms $= 8 \times 10^{-3}$ s;
$\dfrac{\Delta I}{t} = \dfrac{4}{8 \times 10^{-3}} = \dfrac{4000}{8} = 500$ A/s; $E = 1.5$ kV $= 1500$ V

Since $E = L\left(\dfrac{\Delta I}{t}\right)$, $L = \dfrac{E}{(\Delta I/t)} = \dfrac{1500}{500} = \mathbf{3\ H}$

Problem 11 An average e.m.f. of 40 V is induced in a coil of inductance 150 mH when a current of 6 A is reversed. Calculate the time taken for the current to reverse.

$E = 40$ V; $L = 150$ mH $= 0.15$ H; Change in current, $\Delta I = 6 - (-6) = 12$ A (since the current is reversed).

Since $E = L\left(\dfrac{\Delta I}{t}\right)$, time $t = \dfrac{L\Delta I}{E} = \dfrac{(0.15)(12)}{40} = \mathbf{0.045s\ or\ 45\ ms}$

Problem 12 Calculate the coil inductance when a current of 4 A in a coil of 800 turns produces a flux of 5 mWb linking with the coil.

For a coil, inductance $L = \dfrac{N\Phi}{I} = \dfrac{(800)(5 \times 10^{-3})}{4} = \mathbf{1\ H}$

Problem 13 When a current of 1.5 A flows in a coil the flux linking with the coil is $90\,\mu$Wb. If the coil inductance is 0.60 H calculate the number of turns of the coil.

For a coil, $L = \dfrac{N\Phi}{I}$. Thus $N = \dfrac{LI}{\Phi} = \dfrac{(0.6)(1.5)}{90 \times 10^{-6}}$ = **10 000 turns**

Problem 14 When carrying a current of 3 A, a coil of 750 turns has a flux of 12 mWb linking with it. Calculate the coil inductance and the e.m.f. induced in the coil when the current collapses to zero in 18 ms.

Coil inductance, $L = \dfrac{N\Phi}{I} = \dfrac{(750)(12 \times 10^{-3})}{3}$ = **3 H**

Induced e.m.f. $E = L\left(\dfrac{\Delta I}{t}\right) = 3\left(\dfrac{3-0}{18 \times 10^{-3}}\right)$ = **500 V**

(Alternatively, $E = N\left(\dfrac{\Delta \Phi}{t}\right) = (750)\left(\dfrac{12 \times 10^{-3}}{18 \times 10^{-3}}\right)$ = **500 V**)

Problem 15 Calculate the mutual inductance between two coils when a current changing at 200 A/s in one coil induces an e.m.f. of 1.5 V in the other.

Induced e.m.f. $E_2 = M\left(\dfrac{\Delta I_1}{t}\right)$, or $1.5 = M(200)$

Thus mutual inductance, $M = \dfrac{1.5}{200}$ = **0.0075 H or 7.5 mH**

Problem 16 The mutual inductance between two coils is 18 mH. Calculate the steady rate of change of current in one coil to induce an e.m.f. of 0.72 V in the other.

Induced e.m.f. $E_2 = M\left(\dfrac{\Delta I_1}{t}\right)$

Hence rate of change of current, $\dfrac{\Delta I_1}{t} = \dfrac{E_2}{M} = \dfrac{0.72}{0.018}$ = **40 A/s**

Problem 17 Two coils have a mutual inductance of 0.2 H. If the current in one coil is changed from 10 A to 4 A in 10 ms, calculate (a) the average induced e.m.f. in the second coil; (b) the change of flux linked with the second coil if it is wound with 500 turns.

(a) Induced e.m.f. $E_2 = M\left(\dfrac{\Delta I_1}{t}\right) = (0.2)\left(\dfrac{10-4}{10 \times 10^{-3}}\right)$ = **120 V**

(b) Induced e.m.f. $E = N\left(\dfrac{\Delta \Phi}{t}\right)$, hence $\Delta \Phi = \dfrac{Et}{N}$

Thus the change of flux, $\Delta \Phi = \dfrac{120(10 \times 10^{-3})}{500}$ = **2.4 mWb**

Problem 18 A transformer has 500 primary turns and 3000 secondary turns. If the primary voltage is 240 V, determine the secondary voltage, assuming an ideal transformer.

For an ideal transformer, voltage ratio = turns ratio

i.e., $\dfrac{V_1}{V_2} = \dfrac{N_1}{N_2}$, hence $\dfrac{240}{V_2} = \dfrac{500}{3000}$

Thus secondary voltage, $V_2 = \dfrac{(3000)(240)}{(500)} = $ **1440 V or 1.44 kV**

Problem 19 An ideal transformer with a turns ratio of $2 : 7$ is fed from a 240 V supply. Determine its output voltage.

A turns ratio of $2 : 7$ means that the transformer has 2 turns on the primary for every 7 turns on the secondary (i.e. a step-up transformer).

Thus, $\dfrac{N_1}{N_2} = \dfrac{2}{7}$

For an ideal transformer, $\dfrac{N_1}{N_2} = \dfrac{V_1}{V_2}$; hence $\dfrac{2}{7} = \dfrac{240}{V_2}$

Thus the secondary voltage $V_2 = \dfrac{(240)(7)}{(2)} = $ **840 V**

Problem 20 An ideal transformer has a turns ratio of $8 : 1$ and the primary current is 3 A when it is supplied at 240 V. Calculate the secondary voltage and current.

A turns ratio of $8 : 1$ means $\dfrac{N_1}{N_2} = \dfrac{8}{1}$, i.e. a step-down transformer.

$\dfrac{N_1}{N_2} = \dfrac{V_1}{V_2}$, or secondary voltage $V_2 = V_1\left(\dfrac{N_2}{N_1}\right) = 240\left(\dfrac{1}{8}\right) = $ **30 volts**

Also $\dfrac{N_1}{N_2} = \dfrac{I_2}{I_1}$; hence secondary current $I_2 = I_1\left(\dfrac{N_1}{N_2}\right) = 3\left(\dfrac{8}{1}\right) = $ **24 A**

Problem 21 An ideal transformer, connected to a 240 V mains, supplies a 12 V, 150 W lamp. Calculate the transformer turns ratio and the current taken from the supply.

$V_1 = 240$ V; $V_2 = 12$ V; $I_2 = \dfrac{P}{V_2} = \dfrac{150}{12} = 12.5$ A

Turns ratio $= \dfrac{N_1}{N_2} = \dfrac{V_1}{V_2} = \dfrac{240}{12} = $ **20**

$\dfrac{V_1}{V_2} = \dfrac{I_2}{I_1}$, from which $I_1 = I_2\left(\dfrac{V_2}{V_1}\right) = 12.5\left(\dfrac{12}{240}\right)$

Hence current taken from the supply, $I_1 = \dfrac{12.5}{20} = $ **0.625 A**

Problem 22 A 12 Ω resistor is connected across the secondary winding of an ideal transformer whose secondary voltage is 120 V. Determine the primary voltage if the supply current is 4 A.

Secondary current $I_2 = \dfrac{V_2}{R_2} = \dfrac{120}{12} = 10$ A

$\dfrac{V_1}{V_2} = \dfrac{I_2}{I_1}$, from which the primary voltage $V_1 = V_2\left(\dfrac{I_2}{I_1}\right) = 120\left(\dfrac{10}{4}\right) = $ **300 volts**

C. FURTHER PROBLEMS ON ELECTROMAGNETIC INDUCTION

(a) SHORT ANSWER PROBLEMS

1 What is electromagnetic induction?

2 State Faraday's laws of electromagnetic induction.

3 State Lenz's law.

4 Explain briefly the principle of the generator.

5 The direction of an induced e.m.f. in a generator may be determined using Fleming's . . . . rule.

6 The e.m.f. E induced in a moving conductor may be calculated using the formula $E = Blv$. Name the quantities represented and their units.

7 What is self inductance? State its symbol.

8 State and define the unit of inductance.

9 When a circuit has an inductance L and the current changes at a rate of $\left(\frac{\Delta I}{t}\right)$ then the induced e.m.f. E is given by $E = \ldots \ldots$ volts.

10 If a current of I A flowing in a coil of N turns produces a flux of Φ Wb, the coil inductance L is given by $L = \ldots \ldots$ henry's.

11 What is mutual inductance? State its symbol.

12 The mutual inductance between two coils is M. The e.m.f. E_2 induced in one coil by a current changing at $\left(\frac{\Delta I_1}{t}\right)$ in the other is given by $E_2 = \ldots \ldots$ volts.

13 Explain briefly how a voltage is induced in the secondary winding of a transformer.

14 Draw the circuit diagram symbol for a transformer.

15 State the relationship between turns and voltage ratios for a transformer.

(b) MULTI-CHOICE PROBLEMS (answers on page 172)

1 A current changing at a rate of 5 A/s in a coil of inductance 5 H induces an e.m.f. of:
   (a) 25 V in the same direction as the applied voltage;
   (b) 1 V in the same direction as the applied voltage;
   (c) 25 V in the opposite direction to the applied voltage;
   (d) 1 V in the opposite direction to the applied voltage.

2 A bar magnet is moved at a steady speed of 1.0 m/s towards a coil of wire which is connected to a centre-zero galvanometer. The magnet is now withdrawn along the same path at 0.5 m/s. The deflection of the galvanometer is in the:
   (a) same direction as previous with the magnitude of the deflection doubled;
   (b) opposite direction as previous with the magnitude of the deflection halved;
   (c) same direction as previous with the magnitude of the deflection halved;
   (d) opposite direction as previous with the magnitude of the deflection doubled.

3 When a magnetic flux of 10 Wb links with a circuit of 20 turns in 2 s, the induced e.m.f. is:
   (a) 1 V; (b) 4 V; (c) 100 V; (d) 400 V.

4 A current of 10 A in a coil of 1000 turns produces a flux of 10 mWb linking with the coil. The coil inductance is:
(a) 10^6 H; (b) 1 H; (c) 1 μH; (d) 1 mH.

5 An e.m.f. of 1 V is induced in a conductor moving at 10 cm/s in a magnetic field of 0.5 T. The effective length of the conductor in the magnetic field is:
(a) 20 cm; (b) 5 m; (c) 20 m; (d) 50 m.

6 Which of the following statements is false?
(a) Fleming's left-hand rule or Lenz's law may be used to determine the direction of an induced e.m.f.
(b) An induced e.m.f. is set up whenever the magnetic field linking that circuit changes.
(c) The direction of an induced e.m.f. is always such as to oppose the effect producing it.
(d) The induced e.m.f. in any circuit is proportional to the rate of change of the magnetic flux linking the circuit.

7 The mutual inductance between two coils, when a current changing at 20 A/s in one coil induces an e.m.f. of 10 mV in the other is:
(a) 0.5 H; (b) 200 mH; (c) 0.5 mH; (d) 2 H.

8 A transformer has 800 primary turns and 100 secondary turns. To obtain 40 V from the secondary winding the voltage applied to the primary winding must be:
(a) 5 V; (b) 320 V; (c) 2.5 V; (d) 20 V.

9 An e.m.f. is induced into a conductor in the direction shown in *Fig 7* when the conductor is moved at a uniform speed in the field between the two magnets. The polarity of the system is:
(a) North pole on right, South pole on left;
(b) North pole on left, South pole on right.

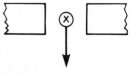

Fig 7

10 An ideal transformer has a turns ratio of 1:5 and is supplied at 200 V when the primary current is 3 A. Which of the following statements is false?
(a) The turns ratio indicates a step-up transformer. (b) The secondary voltage is 40 V. (c) The secondary current is 15 A. (d) The transformer rating is 0.6 kV A. (e) The secondary voltage is 1 kV. (f) The secondary current is 0.6 A.

(c) CONVENTIONAL PROBLEMS

1 A conductor of length 15 cm is moved at 750 mm/s at right-angles to a uniform flux density of 1.2 T. Determine the e.m.f. induced in the conductor.   [0.135 V]

2 Find the speed that a conductor of length 120 mm must be moved at right angles to a magnetic field of flux density 0.6 T to induce in it an e.m.f. of 1.8 V.   [25 m/s]

3 A 25 cm long conductor moves at a uniform speed of 8 m/s through a uniform magnetic field of flux density 1.2 T. Determine the current flowing in the conductor when (a) its ends are open-circuited; (b) its ends are connected to a load of 15 ohms resistance.   [(a) 0; (b) 0.16 A]

4 A straight conductor 500 mm long is moved with constant velocity at right-angles both to its length and to a uniform magnetic field. Given that the e.m.f. induced in the conductor is 2.5 V and the velocity is 5 m/s, calculate the flux density of the

magnetic field. If the conductor forms part of a closed circuit of total resistance 5 ohms, calculate the force on the conductor. [1 T; 0.25 N]

5 A car is travelling at 80 km/h. Assuming the back axle of the car is 1.76 m in length and the vertical component of the earth's magnetic field is 40 μT, find the e.m.f. generated in the axle due to motion. [1.56 mV]

6 A conductor moves with a velocity of 20 m/s at an angle of (a) 90°; (b) 45°; (c) 30°, to a magnetic field produced between two square-faced poles of side length 2.5 cm. If the flux on the pole face is 60 mWb, find the magnitude of the induced e.m.f. in each case. [(a) 48 V; (b) 33.9 V; (c) 24 V]

7 Find the e.m.f. induced in a coil of 200 turns when there is a change of flux of 30 mWb linking with it in 40 ms. [150 V]

8 An e.m.f. of 25 V is induced in a coil of 300 turns when the flux linking with it changes by 12 mWb. Find the time, in milliseconds, in which the flux makes the change. [144 ms]

9 An ignition coil having 10 000 turns has an e.m.f. of 8 kV induced in it. What rate of change of flux is required for this to happen? [0.8 Wb/s]

10 A flux of 0.35 mWb passing through a 125-turn coil is reversed in 25 ms. Find the average e.m.f. induced. [3.5 V]

11 Calculate the e.m.f. induced in a coil of inductance 6 H by a current changing at a rate of 15 A/s. [90 V]

12 An e.m.f. of 2 kV is induced in a coil when a current of 5 A collapses uniformly to zero in 10 ms. Determine the inductance of the coil. [4 H]

13 An average e.m.f. of 50 V is induced in a coil of inductance 160 mH when a current of 7.5 A is reversed. Calculate the time taken for the current to reverse. [48 ms]

14 A coil of 2500 turns has a flux of 10 mWb linking with it when carrying a current of 2 A. Calculate the coil inductance and the e.m.f. induced in the coil when the current collapses to zero in 20 ms. [12.5 H; 1.25 kV]

15 Calculate the coil inductance when a current of 5 A in a coil of 1000 turns produces a flux of 8 mWb linking with the coil. [1.6 H]

16 A coil is wound with 600 turns and has a self inductance of 2.5 H. What current must flow to set up a flux of 20 mWb? [4.8 A]

17 When a current of 2 A flows in a coil, the flux linking with the coil is 80 μWb. If the coil inductance is 0.5 H, calculate the number of turns of the coil. [12 500]

18 A coil of 1200 turns has a flux of 15 mWb linking with it when carrying a current of 4 A. Calculate the coil inductance and the e.m.f. induced in the coil when the current collapses to zero in 25 ms. [4.5 H; 720 V]

19 A coil has 300 turns and an inductance of 4.5 mH. How many turns would be needed to produce a 0.72 mH coil assuming the same core is used? [120 turns]

20 A steady current of 5 A when flowing in a coil of 1000 turns produces a magnetic flux of 500 μWb. Calculate the inductance of the coil. The current of 5 A is then reversed in 12.5 ms. Calculate the e.m.f. induced in the coil. [0.1 H; 80 V]

21 The mutual inductance between two coils is 150 mH. Find the e.m.f. induced in one coil when the current in the other is increasing at the rate of 30 A/s. [4.5 V]

41

22 Determine the mutual inductance between two coils when a current changing at 50 A/s in one coil induces an e.m.f. of 80 mV in the other. [1.6 mH]

23 Two coils have a mutual inductance of 0.75 H. Calculate the e.m.f. induced in one coil when a current of 2.5 A in the other coil is reversed in 15 ms. [250 V]

24 The mutual inductance between two coils is 240 mH. If the current in one coil changes from 15 A to 6 A in 12 ms, calculate (a) the average e.m.f. induced in the other coil; (b) the change of flux linked with the other coil if it is wound with 400 turns. [(a) 180 V; (b) 5.4 mWb]

25 A mutual inductance of 0.6 H exists between two coils. If a current of 6 A in one coil is reversed in 0.8 s calculate (a) the average e.m.f. induced in the other coil; (b) the number of turns on the other coil if the flux change linking with the other coil is 5 mWb. [(a) 0.9 V; (b) 144]

26 A transformer has 800 primary turns and 2000 secondary turns. If the primary voltage is 160 V, determine the secondary voltage assuming an ideal transformer. [400 V]

27 An ideal transformer with a turns ratio of 3 : 8 is fed from a 240 V supply. Determine its output voltage. [640 V]

28 An ideal transformer has a turns ratio of 12 : 1 and is supplied at 192 V. Calculate the secondary voltage. [16 V]

29 A transformer primary winding connected across a 415 V supply has 750 turns. Determine how many turns must be wound on the secondary side if an output of 1.66 kV is required. [3000 turns]

30 (a) Describe the transformer principle of operation.
   (b) An ideal transformer has a turns ratio of 12 : 1 and is supplied at 180 V when the primary current is 4 A. Calculate the secondary voltage and current.
   [15 V; 48 A]

31 A step-down transformer having a turns ratio of 20 : 1 has a primary voltage of 4 kV and a load of 10 kW. Neglecting losses, calculate the value of the secondary current. [50 A]

32 A transformer has a primary-to-secondary turns ratio of 1 : 15. Calculate the primary voltage necessary to supply a 240 V load. If the load current is 3 A, determine the primary current. Neglect any losses. [16 V; 45 A]

4 Alternating voltages and currents

A. MAIN POINTS CONCERNED WITH ALTERNATING VOLTAGES AND CURRENTS

1 Electricity is produced by generators at power stations and then distributed by a vast network of transmission lines (called the National Grid system) to industry and for domestic use. It is easier and cheaper to generate **alternating current** (ac) than direct current (dc) and ac is more conveniently distributed than dc since voltage can be readily altered using transformers. Whenever dc is needed in preference to ac, devices called rectifiers are used for conversion.

2 Let a single turn coil be free to rotate at constant angular velocity ω symmetrically between the poles of a magnet system as shown in *Fig 1*. An emf is generated in

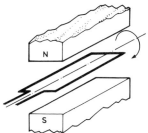

coil (from Faraday's Law) which varies in magnitude and reversed its direction at regular intervals. The reason for this is shown in *Fig 2*. In positions (a), (e) and (i) the conductors of the loop are effectively moving along the magnetic field, no flux is cut and hence no emf is induced. In position (c) maximum flux is cut and hence maximum emf is induced. In position (g), maximum flux is cut and hence maximum emf is again induced.

Fig 1

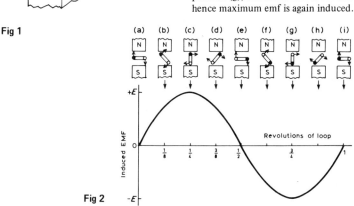

Fig 2

43

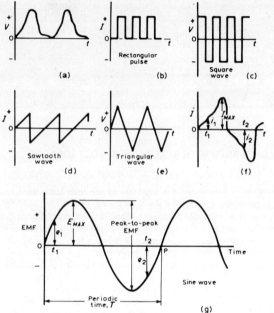

Fig 3

However, using Fleming's right-hand rule, the induced emf is in the opposite direction to that in position (c) and is thus shown as $-E$. In positions (b), (d), (f) and (h) some flux is cut and hence some emf is induced. If all such positions of the coil are considered, in one revolution of the coil, one cycle of alternating emf is produced as shown. This is the principle of operation of the **ac generator** (i.e. the **alternator**).

3   If values of quantities which vary with time t are plotted to a base of time, the resulting graph is called a **waveform**. Some typical waveforms are shown in *Fig 3*. Waveforms (a) and (b) are **unidirectional waveforms**, for, although they vary considerably with time, they flow in one direction only (i.e. they do not cross the time axis and become negative). Waveforms (c) to (g) are called **alternating waveforms** since their quantities are continually changing in direction (i.e. alternately positive and negative).

4   A waveform of the type shown in *Fig 3(g)* is called a **sine wave**. It is the shape of the waveform of emf produced by an alternator and thus the mains electricity supply is of 'sinusoidal' form.

5   One complete series of values is called a **cycle** (i.e. from 0 to P in *Fig 3(g)*).

6   The time taken for an alternating quantity to complete one cycle is called the **period** or the **periodic time**, T, of the waveform.

7   The number of cycles completed in one second is called the **frequency**, f, of the supply and is measured in **hertz**, Hz. The standard frequency of the electricity supply in Great Britain is 50 Hz.

$$T = \frac{1}{f} \quad \text{or} \quad f = \frac{1}{T} . \quad \textit{(See Problems 1 to 3)}.$$

8 **Instantaneous values** are the values of the alternating quantities at any instant of time. They are represented by small letters, i, v, e etc., (see *Figs 3(f)* and *(g)*).

9 The largest value reached in a half cycle is called the **peak value** or the **maximum value** or the **crest value** or the **amplitude** of the waveform. Such values are represented by V_{MAX}, I_{MAX} etc. (see *Figs 3(f)* and *(g)*). A **peak-to-peak** value of emf is shown in *Fig 3(g)* and is the difference between the maximum and minimum values in a cycle.

10 The **average or mean value** of a symmetrical alternating quantity, (such as a sine wave), is the average value measured over a half cycle, (since over a complete cycle the average value is zero).

Average or mean value $= \dfrac{\text{area under the curve}}{\text{length of base}}$

The area under the curve is found by approximate methods such as the trapezoidal rule, the mid-ordinate rule or Simpson's rule. Average values are represented by V_{AV}, I_{AV}, etc.

For a sine wave, average value $= 0.637 \times$ maximum value (i.e. $2/\pi \times$ maximum value)

11 The **effective value** of an alternating current is that current which will produce the same heating effect as an equivalent direct current. The effective value is called the **root mean square (rms) value** and whenever an alternating quantity is given, it is assumed to be the rms value. For example, the domestic mains supply in Great Britain is 240 V and is assumed to mean '240 V rms'. The symbols used for rms values are I, V, E, etc. For a non-sinusoidal waveform as shown in *Fig 4* the rms value is given by:

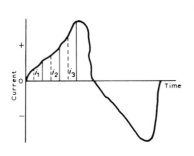

$$I = \sqrt{\left[\frac{i_1{}^2 + i_2{}^2 + \ldots + i_n{}^2}{n}\right]}$$

where n is the number of intervals used.

For a sine wave, rms value $=$ 0.707 $\times$ maximum value (i.e. $1/\sqrt{2} \times$ maximum value).

Fig 4

12 (a) Form factor $= \dfrac{\text{rms value}}{\text{average value}}$. For a sine wave, form factor $= 1.11$.

   (b) Peak factor $= \dfrac{\text{maximum value}}{\text{rms value}}$. For a sine wave, peak factor $= 1.41$.

The values of form and peak factors give an indication of the shape of waveforms. (See *Problems 4 to 8*).

B. WORKED PROBLEMS ON ALTERNATING VOLTAGES AND CURRENTS

Problem 1 Determine the periodic time for frequencies of (a) 50 Hz and (b) 20 kHz.

(a) Periodic time $T = \dfrac{1}{f} = \dfrac{1}{50}$ = 0.02 s or 20 ms

(b) Periodic time $T = \dfrac{1}{f} = \dfrac{1}{20\ 000}$ = 0.000 05 s or 50 μs

Problem 2 Determine the frequencies for periodic times of (a) 4 ms, (b) 4 μs.

(a) Frequency $f = \dfrac{1}{T} = \dfrac{1}{4 \times 10^{-3}} = \dfrac{1000}{4}$ = 250 Hz

(b) Frequency $f = \dfrac{1}{T} = \dfrac{1}{4 \times 10^{-6}} = \dfrac{1\ 000\ 000}{4}$ = 250 000 Hz or 250 kHz or 0.25 MHz

Problem 3 An alternating current completes 5 cycles in 8 ms. What is its frequency?

Time for 1 cycle $= \dfrac{8}{5}$ ms = 1.6 ms = periodic time T.

Frequency $f = \dfrac{1}{T} = \dfrac{1}{1.6 \times 10^{-3}} = \dfrac{1000}{1.6} = \dfrac{10\ 000}{16}$ = 625 Hz

Problem 4 For the periodic waveforms shown in *Fig 5* determine for each: (i) frequency; (ii) average value over half a cycle; (iii) rms value; (iv) form factor; and (v) peak factor.

(a) *Triangular waveform (Fig 5(a))*

(i) Time for 1 complete cycle = 20 ms = periodic time, T.

Hence frequency $f = \dfrac{1}{T} = \dfrac{1}{20 \times 10^{-3}} = \dfrac{1000}{20}$ = 50 Hz

(ii) Area under the triangular waveform for a half cycle

$= \dfrac{1}{2} \times$ base $\times$ height

46

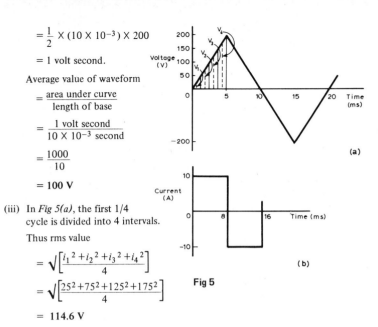

$$= \frac{1}{2} \times (10 \times 10^{-3}) \times 200$$

$$= 1 \text{ volt second.}$$

Average value of waveform

$$= \frac{\text{area under curve}}{\text{length of base}}$$

$$= \frac{1 \text{ volt second}}{10 \times 10^{-3} \text{ second}}$$

$$= \frac{1000}{10}$$

$$= 100 \text{ V}$$

(iii) In *Fig 5(a)*, the first 1/4 cycle is divided into 4 intervals.

Thus rms value

$$= \sqrt{\left[\frac{i_1{}^2 + i_2{}^2 + i_3{}^2 + i_4{}^2}{4}\right]}$$

$$= \sqrt{\left[\frac{25^2 + 75^2 + 125^2 + 175^2}{4}\right]}$$

$$= 114.6 \text{ V}$$

Fig 5

(Note that the greater the number of intervals chosen, the greater the accuracy of the result. For example, if twice the number of ordinates as that chosen above are used, the rms value is found to be 115.6 V)

(iv) Form factor $= \dfrac{\text{rms value}}{\text{average value}} = \dfrac{114.6}{100} = 1.15$

(v) Peak factor $= \dfrac{\text{maximum value}}{\text{rms value}} = \dfrac{200}{114.6} = 1.75$

(b) *Rectangular waveform (Fig 5(b))*

(i) Time for 1 complete cycle = 16 ms = periodic time, T.

Hence frequency, $f = \dfrac{1}{T} = \dfrac{1}{16 \times 10^{-3}} = \dfrac{1000}{16} = 62.5 \text{ Hz}$

(ii) Average value over half a cycle $= \dfrac{\text{area under curve}}{\text{length of base}}$

$$= \frac{10 \times (8 \times 10^{-3})}{8 \times 10^{-3}} = 10 \text{ A}$$

(iii) The rms value $= \sqrt{\left[\dfrac{i_1{}^2 + i_2{}^2 + \ldots + i_n{}^2}{n}\right]} = 10 \text{ A}$,

however many intervals are chosen, since the waveform is rectangular.

(iv) Form factor $= \dfrac{\text{rms value}}{\text{average value}} = \dfrac{10}{10} = 1$

(v) Peak factor $= \dfrac{\text{maximum value}}{\text{rms value}} = \dfrac{10}{10} = 1$

Problem 5 The following table gives the corresponding values of current and time for a half cycle of alternating current.

time t (ms)	0	0.5	1.0	1.5	2.0	2.5	3.0	3.5	4.0	4.5	5.0
current i (A)	0	7	14	23	40	56	68	76	60	5	0

Assuming the negative half cycle is identical in shape to the positive half cycle, plot the waveform and find (a) the frequency of the supply, (b) the instantaneous values of current after 1.25 ms and 3.8 ms, (c) the peak or maximum value, (d) the mean or average value, and (e) the rms value of the waveform.

The half cycle of alternating current is shown plotted in *Fig 6*.

(a) Time for a half cycle = 5 ms. Hence the time for 1 cycle, i.e. the periodic time, $T = 10$ ms or 0.01 s.

Frequency, $f = \dfrac{1}{T} = \dfrac{1}{0.01} = \textbf{100 Hz}$.

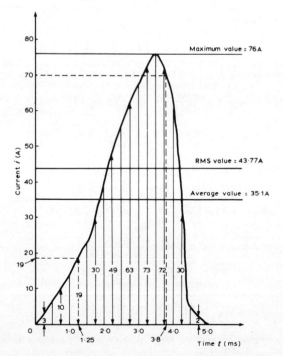

Fig 6

48

(b) Instantaneous value of current after 1.25 ms is **19 A**, from *Fig 6*.
 Instantaneous value of current after 3.8 ms is **70 A**, from *Fig 6*.
(c) Peak or maximum value = **76A**

(d) Mean or average value = $\dfrac{\text{area under curve}}{\text{length of base}}$

 Using the mid-ordinate rule with 10 intervals, each of width 0.5 ms gives:

 Area under curve = $(0.5 \times 10^{-3})[3+10+19+30+49+63+73+72+30+2]$

 (see *Fig 6*)

 $= (0.5 \times 10^{-3})(351)$

 Hence mean or average value = $\dfrac{(0.5 \times 10^{-3})(351)}{5 \times 10^{-3}}$ = **35.1 A**

(e) rms value = $\sqrt{\left[\dfrac{3^2+10^2+19^2+30^2+49^2+63^2+73^2+72^2+30^2+2^2}{10}\right]}$

 $= \sqrt{\left[\dfrac{19\,157}{10}\right]}$ = **43.8 A**

Problem 6 Calculate the rms value of a sinusoidal current of maximum value 20 A.

For a sine wave,
rms value = 0.707 × maximum value
 = 0.707 × 20 = **14.14 A**

Problem 7 Determine the peak and mean values for a 240 V mains supply.

For a sine wave, rms value of voltage $V = 0.707 \times V_{MAX}$
A 240 V mains supply means that 240 V is the rms value.

Hence $\quad V_{MAX} = \dfrac{V}{0.707} = \dfrac{240}{0.707}$ = **339.5 V = peak value**

Mean value V_{AV} = $0.637\ V_{MAX}$ = 0.637 × 339.5 = **216.3 V**

Problem 8 A supply voltage has a mean value of 150 V. Determine its maximum value and its rms value.

For a sine wave, mean value = 0.637 × maximum value.

Hence maximum value = $\dfrac{\text{mean value}}{0.637} = \dfrac{150}{0.637}$ = **235.5 V**

rms value = 0.707 × maximum value = 0.707 × 235.5 = **166.5 V**

C. FURTHER PROBLEMS ON ALTERNATING VOLTAGES AND CURRENTS

(a) SHORT ANSWER PROBLEMS

1   Briefly explain the principle of the simple alternator.

2   What is the difference between an alternating and a unidirectional waveform.

3   What is meant by (a) waveform; (b) cycle.

4   The time to complete one cycle of a waveform is called the . . . .

5   What is frequency? Name its unit.

6   The mains supply voltage has a special shape of waveform called a . . . .

7   Define peak value.

8   What is meant by the rms value?

9   The domestic mains electricity supply voltage in Great Britain is . . . .

10 What is the mean value of a sinusoidal alternating e.m.f. which has a maximum value of 100 V?

11 The effective value of a sinusoidal waveform is (. . . . . × maximum value).

(b) MULTI-CHOICE PROBLEMS (answers on page 172)

1   The value of an alternating current at any given instant is (a) a maximum value; (b) a peak value; (c) an instantaneous value; (d) an rms value.

2   An alternating current completes 100 cycles in 0.1 s. Its frequency is (a) 20 Hz; (b) 100 Hz; (c) 0.002 Hz; (d) 1 kHz.

3   In *Fig 7*, at the instant shown the generated e.m.f. will be (a) zero; (b) an rms value; (c) an average value; (d) a maximum value.

4   The supply of electrical energy for a consumer is usually by a.c. because:
    (a) transmission and distribution are more easily effected;
    (b) it is most suitable for variable speed motors;
    (c) the volt drop in cables is minimal;
    (d) cable power losses are negligible.

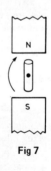

Fig 7

5   Which of the following statements is false?
    (a) It is cheaper to use a.c. than d.c.
    (b) Distribution of a.c. is more convenient than with d.c. since voltages may be readily altered using transformers.
    (c) An alternator is an a.c. generator.
    (d) A rectifier changes d.c. into a.c.

6   An alternating voltage of maximum value 100 V is applied to a lamp. Which of the following direct voltages, if applied to the lamp, would cause the lamp to light with the same brilliance?
    (a) 100 V; (b) 63.7 V; (c) 70.7 V; (d) 141.4 V.

7 The value normally stated when referring to alternating currents and voltages is the:
(a) instantaneous value; (b) rms value; (c) average value; (d) peak value.

8 State which of the following is false. For a sine wave:
(a) the peak factor is 1.414.
(b) the rms value is 0.707 × peak value.
(c) the average value is 0.637 × rms value.
(d) the form factor is 1.11.

9 An a.c. supply is 70.7 V, 50 Hz. Which of the following statements is false?
(a) The periodic time is 20 ms.
(b) The peak value of the voltage is 70.7 V.
(c) The rms value of the voltage is 70.7 V.
(d) The peak value of the voltage is 100 V.

(c) CONVENTIONAL PROBLEMS

Frequency and periodic time

1 Determine the periodic time for the following frequencies:
(a) 2.5 Hz; (b) 100 Hz; (c) 40 kHz.   [(a) 0.4 s; (b) 10 ms; (c) 25 μs]

2 Calculate the frequency for the following periodic times:
(a) 5 ms; (b) 50 μs; (c) 0.2 s.   [(a) 0.2 kHz; (b) 20 kHz; (c) 5 Hz]

3 An alternating current completes 4 cycles in 5 ms. What is the frequency?
[800 Hz]

a.c. values of non-sinusoidal waveforms

4 An alternating current varies with time over half a cycle as follows:

Current (A)	0	0.7	2.0	4.2	8.4	8.2	2.5	1.0	0.4	0.2	0
time (ms)	0	1	2	3	4	5	6	7	8	9	10

The negative half cycle is similar. Plot the curve and determine:
(a) the frequency; (b) the instantaneous values at 3.4 ms and 5.8 ms; (c) its mean
value and (d) its rms value.   [(a) 50 Hz; (b) 5.5 A, 3.4 A; (c) 2.8 A; (d) 4.0 A]

5 For the waveforms shown in *Fig 8* determine for each (i) the frequency; (ii) the
average value over half a cycle; (iii) the rms value; (iv) the form factor; (v) the peak
factor.

⎡ (a) (i) 100 Hz; (ii) 2.50A; (iii) 2.88A; (iv) 1.15; (v) 1.74 ⎤
| (b) (i) 250 Hz; (ii) 20 V; (iii) 20 V; (iv) 1.0; (v) 1.0 |
| (c) (i) 125 Hz; (ii) 18 A; (iii) 19.56 A; (iv) 1.09; (v) 1.23 |
⎣ (d) (i) 250 Hz; (ii) 25 V; (iii) 50 V; (iv) 2.0; (v) 2.0 ⎦

6 An alternating voltage is triangular in shape, rising at a constant rate to a maximum
of 300 V in 8 ms and then falling to zero at a constant rate in 4 ms. The negative
half cycle is identical in shape to the positive half cycle. Calculate (a) the mean
voltage over half a cycle, and (b) the rms voltage.   [(a) 150 V; (b) 170 V]

7 An alternating e.m.f. varies with time over half a cycle as follows:

E.m.f. (V)	0	45	80	155	215	320	210	95	0
time (ms)	0	1.5	3.0	4.5	6.0	7.5	9.0	10.5	12.0

The negative half cycle is identical in shape to the positive half cycle. Plot the waveform and determine (a) the periodic time and frequency; (b) the instantaneous value

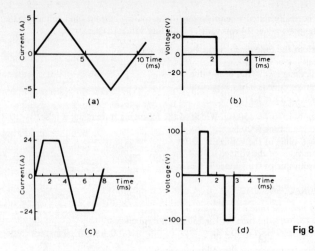

Fig 8

of voltage at 3.75 ms, (c) the times when the voltage is 125 V; (d) the mean value, and (e) the rms value.

[(a) 24 ms, 41.67 Hz; (b) 115 V; (c) 4 ms and 10.1 ms; (d) 142 V; (e) 171 V]

a.c. values of sinusoidal waveforms

8 Calculate the rms value of a sinusoidal curve of maximum value 300 V.   [212.1 V]

9 Find the peak and mean values for a 200 V mains supply.   [282.9 V; 180.2 V]

10 Plot a sine wave of peak value 10.0 A. Show that the average value of the waveform is 6.37 A over half a cycle, and that the rms value is 7.07 A.

11 A sinusoidal voltage has a maximum value of 120 V. Calculate its rms and average values.   [84.8 V; 76.4 V]

12 A sinusoidal current has a mean value of 15.0 A. Determine its maximum and rms values.   [23.55 A; 16.65 A]

5 Electrical measuring instruments and measurements

A MAIN POINTS CONCERNING ELECTRICAL MEASURING INSTRUMENTS AND MEASUREMENTS

1 Tests and measurements are important in designing, evaluating, maintaining and servicing electrical circuits and equipment. In order to detect electrical quantities such as current, voltage, resistance or power, it is necessary to transform an electrical quantity or condition into a visible indication. This is done with the aid of instruments (or meters) that indicate the magnitude of quantities either by the position of a pointer moving over a graduated scale (called an analogue instrument) or in the form of a decimal number (called a digital instrument).

2 All analogue electrical indicating instruments require three essential devices:
   (a) A **deflecting or operating device**. A mechanical force is produced by the current or voltage which causes the pointer to deflect from its zero position.
   (b) A **controlling device**. The controlling force acts in opposition to the deflecting force and ensures that the deflection shown on the meter is always the same for a given measured quantity. It also prevents the pointer always going to the maximum deflection. There are two main types of controlling device – spring control and gravity control.
   (c) A **damping device**. The damping force ensures that the pointer comes to rest in its final position quickly and without undue oscillation. There are three main types of damping used – eddy-current damping, air-friction damping and fluid-friction damping.

3 There are basically two types of scale – linear and non-linear. A **linear scale** is shown in *Fig 1(a)*, where the divisions or graduations are evenly spaced. The voltmeter shown has a range 0–100 V, i.e. a full-scale deflection (f.s.d.) of 100 V. A **non-linear scale** is shown in *Figure 1(b)*. The scale is cramped at the beginning and the graduations are uneven throughout the range. The ammeter shown has a f.s.d. of 10 A.

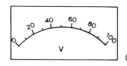
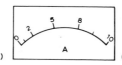

(a)   (b)   **Fig 1**

4 Comparison of the moving coil, moving iron and moving coil rectifier instruments

Type of instrument	Moving coil	Moving iron	Moving coil rectifier
Suitable for measuring	Direct current and voltage	Direct and alternating current and voltage (reading in rms value)	Alternating current and voltage (reads average value but scale is adjusted to give rms value for sinsuoidal waveforms)
Scale	Linear	Non-linear	Linear
Method of control	Hairsprings	Hairsprings	Hairsprings
Method of damping	Eddy current	Air	Eddy current
Frequency limits	–	20–200 Hz	20–100 kHz
Advantages	1 Linear scale 2 High sensitivity 3 Well shielded from stray magnetic fields 4 Lower power consumption	1 Robust construction 2 Relatively cheap 3 Measures dc and ac 4 In frequency range 20–100 Hz reads rms correctly, regardless of supply waveform	1 Linear scale 2 High sensitivity 3 Well shielded from stray magnetic fields 4 Low power consumption 5 Good frequency range
Disadvantages	1 Only suitable for dc 2 More expensive than moving iron type 3 Easily damaged	1 Non-linear scale 2 Affected by stray magnetic fields 3 Hysteresis errors in dc circuits 4 Liable to temperature errors 5 Due to the inductance of the solenoid, readings can be affected by variation of frequency	1 More expensive than moving iron type 2 Errors caused when supply is non-sinusoidal

(For the principle of operation of a moving-coil instrument, see *Problem 9*, chapter 2, page 26, and for a moving-iron instrument, see *Worked Problem 1* of this chapter.)

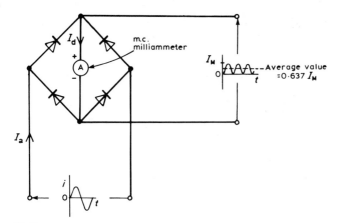

Fig 2

5 A moving-coil instrument, which measures only d.c., may be used in conjunction with a bridge rectifier circuit as shown in *Fig 2* to provide an indication of alternating currents and voltages (see chapter 4). The average value of the full-wave rectified current is $0.637 I_M$. However, a meter being used to measure a.c. is usually calibrated in r.m.s. values. For sinusoidal quantities the indication is

$$\frac{0.707 I_M}{0.637 I_M}$$

i.e. 1.11 times the mean value. Rectifier instruments have scaled calibrated in r.m.s. quantities and it is assumed by the manufacturer that the a.c. is sinusoidal.

6 An **ammeter**, which measures current, has a low resistance (ideally zero) and must be connected in series with the circuit.

7 A **voltmeter**, which measures p.d., has a high resistance (ideally infinite) and must be connected in parallel with the part of the circuit whose p.d. is required.

8 There is no difference between the basic instrument used to measure current and voltage since both use a milliammeter as their basic part. This is a sensitive instrument which gives f.s.d. for currents of only a few milliamperes. When an ammeter is required to measure currents of larger magnitude, a proportion of the current is diverted through a low-value resistance connected in parallel with the meter. Such a diverting resistor is called a **shunt**.

From *Figure 3(a)*, $V_{PQ} = V_{RS}$. Hence $I_a r_a = I_S R_S$

Thus the value of the shunt, $R_S = \dfrac{I_a r_a}{I_s}$ ohms

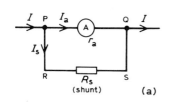

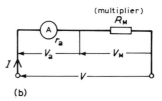

Fig 3

The milliammeter is converted into a voltmeter by connecting a high resistance (called a **multiplier**) in series with it as shown in *Fig 3(b)*. From *Fig 3(b)*.

$$V = V_a + V_M = Ir_a + IR_M$$

Thus the value of the multiplier $R_M = \dfrac{V - Ir_a}{I}$ ohms.

(See *Problems 2 and 3*).

9 An **ohmmeter** is an instrument for measuring electrical resistance.
A simple ohmmeter circuit is shown in *Fig 4(a)*. Unlike the ammeter or voltmeter, the ohmmeter circuit does not receive the energy necessary for its operation from the circuit under test. In the ohmmeter this energy is supplied by a self-contained source of voltage, such as a battery. Initially, terminals XX are short-circuited and R adjusted to give f.s.d. on the milliammeter.
If current I is at a maximum value and voltage E is constant, then resistance $R = E/I$

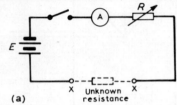

(a)

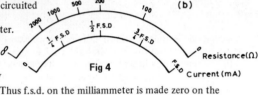

(b)

Fig 4

is at a minimum value. Thus f.s.d. on the milliammeter is made zero on the resistance scale. When terminals XX are open circuited no current flows and $R (= E/0)$ is infinity, ∞.

The milliammeter can thus be calibrated directly in ohms. A cramped (non-linear) scale results and is 'back to front', as shown in *Fig 4(b)*. When calibrated, an unknown resistance is placed between terminals XX and its value determined from the position of the pointer on the scale. An ohmmeter designed for measuring low values of resistance is called a **continuity tester**. An ohmmeter designed for measuring high values of resistance (i.e. megohms) is called an **insulation resistance tester** (e.g. 'Megger').

10 Instruments are manufactured that combine a moving-coil meter with a number of shunts and series multipliers, to provide a range of readings on a single scale graduated to read current and voltage. If a battery is incorporated then resistance can also be measured. Such instruments are called **multimeters** or **universal instruments** or **multirange instruments**. An 'Avometer' is a typical example. A particular range may be selected either by the use of separate terminals or by a selector switch. Only one measurement can be performed at one time. Often such instruments can be used in a.c. as well as d.c. circuits when a rectifier is incorporated in the instrument.

11 A **wattmeter** is an instrument for measuring electrical power in a circuit. *Fig 5* shows typical connections of a wattmeter used for measuring power supplied to a load. The instrument has two coils:
(i) a current coil, which is connected in series with the load, like an ammeter, and
(ii) a voltage coil, which is connected in parallel with the load, like a voltmeter.
(See *Problems 4 and 5*).

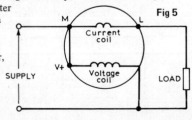

Fig 5

56

12 The **cathode ray oscilloscope (c.r.o.)** may be used in the observation of waveforms and for the measurement of voltage, current, frequency, phase and periodic time. For examining periodic waveforms the electron beam is deflected horizontally (i.e. in the X direction) by a sawtooth generator acting as a timebase. The signal to be examined is applied to the vertical deflection system (Y direction) usually after amplification.

Oscilloscopes normally have a transparent grid of 10 mm by 10 mm squares in front of the screen, called a graticule. Among the timebase controls is a 'variable' switch which gives the sweep speed as time per centimetre. This may be in s/cm, ms/cm or μs/cm, a large number of switch positions being available. Also on the front panel of a c.r.o. is a Y amplifier switch marked in volts per centimetre, with a large number of available switch positions.

(i) With **direct voltage measurements**, only the Y amplifier 'volts/cm' switch on the c.r.o. is used. With no voltage applied to the Y plates the position of the spot trace on the screen is noted. When a direct voltage is applied to the Y plates the new position of the spot trace is an indication of the magnitude of the voltage. For example, in *Fig 6(a)*, with no voltage applied to the Y plates, the spot trace is in the centre of the screen (initial position) and then the spot trace moves 2.5 cm to the final position shown, on application of a d.c. voltage. With the 'volts/cm' switch on 10 volts/cm the magnitude of the direct voltage is 2.5 cm × 10 volts/cm, i.e. 25 volts.

(ii) With **alternating voltage measurement**, let a sinusoidal waveform be displayed on a c.r.o. screen as shown in *Fig 6(b)*. If the s/cm switch is on, say, 5 ms/cm then the **periodic time** T of the sinewave is 5 ms/cm × cm, i.e. **20 ms or 0.02 s.**

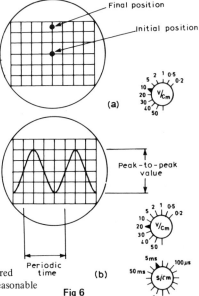

Since frequency $f = \dfrac{1}{T}$,

frequency $= \dfrac{1}{0.02} = 50$ **Hz**

If the 'volts/cm' switch is on, say, 20 volts/cm then the **amplitude or peak value** of the sinewave shown is 20 volts/cm × 2 cm, i.e. **40 V.**

Since r.m.s. voltage $= \dfrac{\text{peak voltage}}{\sqrt{2}}$

(see Chapter 4),

r.m.s. voltage $= \dfrac{40}{\sqrt{2}} = $ **28.28 volts**

Double beam oscilloscopes are useful whenever two signals are to be compared simultaneously. The c.r.o. demands reasonable skill in adjustment and use. However

Fig 6

its greatest advantage is in observing the shape of a waveform – a feature not possessed by other measuring instruments. (See *Problems 6 to 10*).

13 An **electronic voltmeter** can be used to measure with accuracy e.m.f. or p.d. from millivolts to kilovolts by incorporating in its design amplifiers and attenuators.

14 A **null method of measurement** is a simple, accurate and widely used method which depends on an instrument reading being adjusted to read zero current only. The method assumes:

(i) if there is any deflection at all, then some current is flowing;

(ii) if there is no deflection, then no current flows (i.e. a null condition).

Hence it is unnecessary for a meter sensing current flow to be calibrated when used in this way. A sensitive milliammeter or microammeter with centre zero position setting is called a **galvanometer**. Two examples where the method is used are in the Wheatstone bridge and in the d.c. potentiometer.

15 *Fig 7* shows a **Wheatstone bridge** circuit which compares an unknown resistance R_x with others of known values, i.e. R_1 and R_2 which have fixed values, and R_3 which is variable. R_3 is varied until zero deflection is obtained on the galvanometer G. No current then flows through the meter, $V_A = V_B$, and the bridge is said to be 'balanced'.

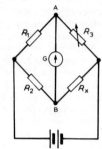

Fig 7

At balance, $R_1 R_x = R_2 R_3$, i.e. $R_x = \dfrac{R_2 R_3}{R_1}$ ohms

(See *Problem 11*)

16 The d.c. **potentiometer** is a null-balance instrument used for determining values of e.m.f.s and p.d.s. by comparison with a known e.m.f. or p.d. In *Fig 8(a)*, using a standard cell of known e.m.f. E_1, the slider S is moved along the slide wire until

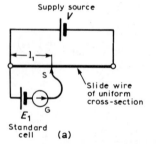
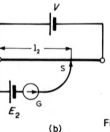

Fig 8

balance is obtained (i.e. the galvanometer deflection is zero), shown as length l_1. The standard cell is now replaced by a cell of unknown e.m.f. E_2 (see *Fig 8(b)*) and again balance is obtained (shown as l_2). Since $E_1 \propto l_1$ and $E_2 \propto l_2$,

then $\dfrac{E_1}{E_2} = \dfrac{l_1}{l_2}$ and $E_2 = E_1 \left(\dfrac{l_2}{l_1}\right)$ volts (see *Problem 12*).

A potentiometer may be arranged as a resistive two-element potential divider in which the division ratio is adjustable to give a simple variable d.c. supply. Such devices may be constructed in the form of a resistive element carrying a sliding contact which is adjusted by a rotary or linear movement of the control knob.

17 The errors most likely to occur in measurements are those due to:

(i) the limitations of the instrument; (ii) the operator; (iii) the instrument disturbing the circuit (see *Problem 13*).

Problem 1 Draw diagrams to represent (a) the attraction type; (b) the repulsion type, of the moving-iron instrument and describe briefly, for each, its principle of operation.

(a) An **attraction type** of moving-iron instrument is shown diagrammatically in *Fig 9(a)*. When current flows in the solenoid, a pivoted soft-iron disc is attracted towards the solenoid and the movement causes a pointer to move across a scale.

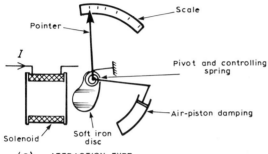

(a)   ATTRACTION TYPE

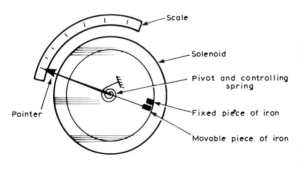

(b)   REPULSION TYPE

Fig 9

(b) In the **repulsion type** moving-iron instrument shown diagrammatically in *Fig 9(b)*, two pieces of iron are placed inside the solenoid, one being fixed, and the other attached to the spindle carrying the pointer. When current passes through the solenoid, the two pieces of iron are magnetised in the same direction and therefore repel each other. The pointer thus moves across the scale. The force moving the pointer is, in each type, proportional to I^2.

Because of this the direction of current does not matter and the moving-iron instrument can be used on d.c. or a.c. The scale, however, is non-linear.

Problem 2 A moving-coil instrument gives a f.s.d. when the current is 40 mA and its resistance is 25 Ω. Calculate the value of the shunt to be connected in parallel with the meter to enable it to be used as an ammeter for measuring currents up to 50 A.

The circuit diagram is shown in *Fig 10*,
where r_a = resistance of instrument
　　　　= 25 Ω;
R_s = resistance of shunt;
I_a = maximum permissible current
　flowing in instrument = 40 mA
　　　　　　　　= 0.04 A;
I_s = current flowing in shunt;
I = total circuit current required to
give f.s.d. = 50 A

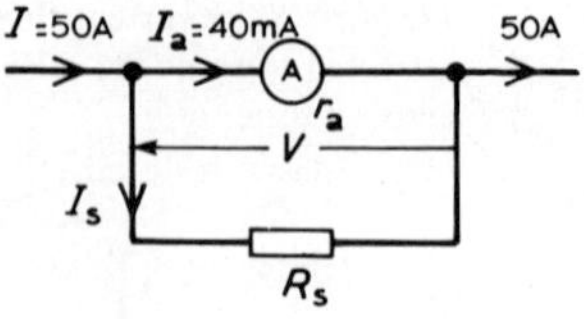

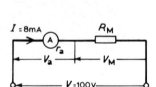

Fig 10

Since $I = I_a + I_s$ then $I_s = I - I_a = 50 - 0.04 = 49.96$ A
　　$V = I_a r_a = I_s R_s$
Hence $R_s = \dfrac{I_a r_a}{I_s} = \dfrac{(0.04)(25)}{49.96} = 0.02002$ Ω $= 20.02$ mΩ

Thus for the moving-coil instrument to be used as an ammeter with a range 0–50 A, a resistance of value 20.02 mΩ needs to be connected in parallel with the instrument.

Problem 3 A moving-coil instrument having a resistance of 10 Ω, gives a f.s.d. when the current is 8 mA. Calculate the value of the multiplier to be connected in series with the instrument so that it can be used as a voltmeter for measuring p.d.s up to 100 V.

The circuit diagram is shown in *Fig 11*,
where r_a = resistance of instrument
　　　　= 10 Ω;
R_M = resistance of multiplier;
I = total permissible instrument
　current = 8 mA = 0.008 A;
V = total p.d. required to give f.s.d.
　　= 100 V.

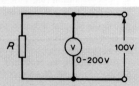

Fig 11

$V = V_a + V_M = I r_a + I R_M$
i.e. $100 = (0.008)(10) + (0.008)R_M$, or $100 - 0.08 = 0.008 R_M$
$R_M = \dfrac{99.92}{0.008} = 12\,490$ Ω $= 12.49$ kΩ

Hence for the moving-coil instrument to be used as a voltmeter with a range 0–100 V, a resistance of value 12.49 kΩ needs to be connected in series with the instrument.

Problem 4 Calculate the power dissipated by the voltmeter and by resistor R in *Fig 12* when (a) $R = 250$ Ω; (b) $R = 2$ MΩ. Assume that the voltmeter sensitivity (sometimes called figure of merit) is 10 kΩ/V.

Fig 12

(a) Resistance of voltmeter, R_v = sensitivity × f.s.d.

Hence, R_v = 10 kΩ/V × 200 V = 2000 kΩ = 2 MΩ

Current flowing in voltmeter, $I_v = \dfrac{V}{R_v} = \dfrac{100}{2 \times 10^6} = 50 \times 10^{-6}$ A.

Power dissipated by voltmeter = $VI_v = (100)(50 \times 10^{-6}) = $ **5 mW**

When $R = 250$ Ω, current in resistor, $I_R = \dfrac{V}{R} = \dfrac{100}{250} = 0.4$ A

Power dissipated in load resistor $R = VI_R = (100)(0.4) = $ **40 W**

Thus the power dissipated in the voltmeter is insignificant in comparison with the power dissipated in the load.

(b) When $R = 2$ MΩ, current in resistor, $I_R = \dfrac{V}{R} = \dfrac{100}{2 \times 10^6} = 50 \times 10^{-6}$ A

Power dissipated in load resistor $R = VI_R = 100 \times 50 \times 10^{-6} = $ **5 mW**

In this case the higher load resistance reduced the power dissipated such that the voltmeter is using as much power as the load.

Problem 5 An ammeter has a f.s.d. of 100 mA and a resistance of 50 Ω. The ammeter is used to measure the current in a load of resistance 500 Ω when the supply voltage is 10 V. Calculate (a) the ammeter reading expected (neglecting its resistance); (b) the actual current in the circuit; (c) the power dissipated in the ammeter; (d) the power dissipated in the load.

From *Fig 13*.

(a) Expected ammeter reading

$= \dfrac{V}{R} = \dfrac{10}{500} = $ **20 mA**

(b) Actual ammeter reading

$= \dfrac{V}{R+r_a} = \dfrac{10}{500+50} = $ **18.18 mA**

Thus the ammeter itself has caused the circuit conditions to change from 20 mA to 18.18 mA.

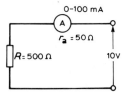

Fig 13

(c) Power dissipated in the ammeter = $I^2R = (18.18 \times 10^{-3})^2(50) = $ **16.53 mW**

(d) Power dissipated in the load resistor = $I^2R = (18.18 \times 10^{-3})^2(500) = $ **165.3 mW**

Problem 6 Describe how a simple c.r.o. is adjusted to give (a) a spot trace; (b) a continuous horizontal trace on the screen, explaining the functions of the various controls.

(a) To obtain a spot trace on a typical c.r.o. screen:

(i)   Switch on the c.r.o.

(ii) Switch the timebase control to off. This control is calibrated in time per centimetres — for example 5 ms/cm or 100 μs/cm. Turning it to zero ensures no signal is applied to the X-plates. The Y-plate input is left open-circuited.

(iii) Set the intensity, X-shift and Y-shift controls to about the mid-range positions.

(iv) A spot trace should now be observed on the screen. If not, adjust either or both of the X and Y-shift controls. The X-shift control varies the position of the spot trace in a horizontal direction whilst the Y-shift control varies its vertical position.

(v)   Use the X and Y-shift controls to bring the spot to the centre of the screen and use the focus control to focus the electron beam into a small circular spot.

(b) To obtain a continuous horizontal trace on the screen the same procedure as in (a) is initially adopted. Then the timebase control is switched to a suitable position, initially the millisecond timebase range, to ensure that the repetition rate of the sawtooth is sufficient for the persistence of the vision time of the screen phosphor to hold a given trace.

Problem 7 For the c.r.o. square voltage waveform shown in *Fig 14* determine (a) the periodic time, (b) the frequency and (c) the peak-to-peak voltage. The 'time/cm' (or timebase control) switch is on 100 μs/cm and the 'volts/cm' (or signal amplitude control) switch is on 20 V/cm.

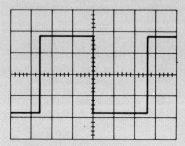

Fig 14

(In *Fig 14 to 17* assume that the squares shown are 1 cm by 1 cm.)

(a) The width of one complete cycle is 5.2 cm.
Hence the periodic time, $T = 5.2$ cm $\times$ 100 $\times$ 10^{-6} s/cm = **0.52 ms**

(b) Frequency, $f = \dfrac{1}{T} = \dfrac{1}{0.52 \times 10^{-3}}$ = **1.92 kHz**

(c) The peak-to-peak height of the display is 3.6 cm, hence the peak-to-peak voltage = 3.6 cm $\times$ 20 V/cm = **72 V**

Problem 8 For the c.r.o. display of a pulse waveform shown in *Fig 15* the 'time/cm' switch is on 50 ms/cm and the 'volts/cm' switch is on 0.2 V/cm. Determine (a) the periodic time; (b) the frequency; (c) the magnitude of the pulse voltage.

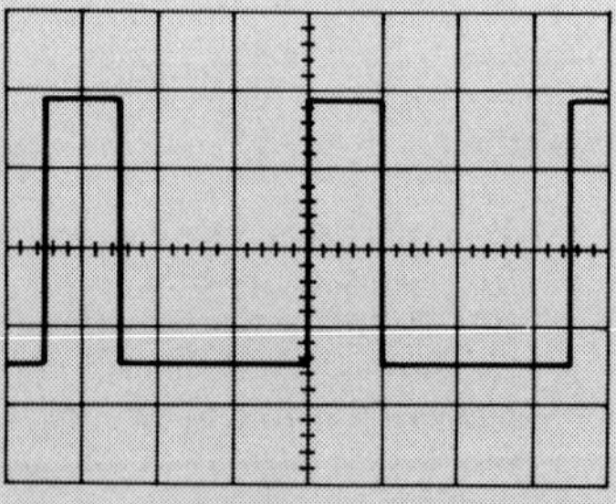

Fig 15

(a) The width of one complete cycle is 3.5 cm.
Hence the periodic time, $T = 3.5$ cm $\times$ 50 ms/cm = **175 ms**

(b) Frequency, $f = \dfrac{1}{T} = \dfrac{1}{0.175} = $ **5.71 Hz**

(c) The height of a pulse is 3.4 cm.

Hence the magnitude of the pulse voltage = 3.4 cm $\times$ 0.2 V/cm = **0.68 V**

Problem 9 A sinusoidal voltage trace displayed by a c.r.o. is shown in *Fig 16*. If the 'time/cm' switch is on 500 μs/cm and the 'volts/cm' switch is on 5 V/cm, find, for the waveform, (a) the frequency, (b) the peak-to-peak voltage; (c) the amplitude; (d) the r.m.s. value.

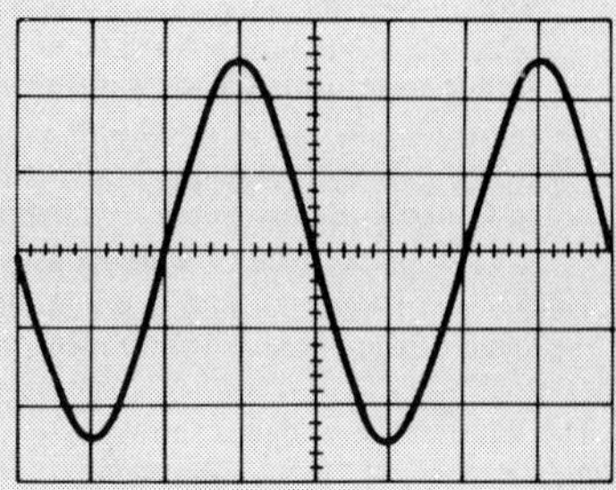

Fig 16

(a) The width of one complete cycle is 4 cm. Hence the periodic time, T is 4 cm $\times$ 500 μs/cm, i.e. 2 ms.

Frequency, $f = \dfrac{1}{T} = \dfrac{1}{2 \times 10^{-3}} = $ **500 Hz**

(b) The peak-to-peak height of the waveform is 5 cm. Hence the peak-to-peak voltage = 5 cm $\times$ 5 V/cm = **25 V**

(c) Amplitude = $\dfrac{1}{2} \times 25$ V = **12.5 V**

(d) The peak value of voltage is the amplitude, i.e. 12.5 V

r.m.s. voltage = $\dfrac{\text{peak voltage}}{\sqrt{2}} = \dfrac{12.5}{\sqrt{2}} = $ **8.84 V**

Problem 10 For the double-beam oscilloscope displays shown in *Fig 17* determine (a) their frequency; (b) their r.m.s. values; (c) their phase difference. The 'time/cm' switch is on 100 μs/cm and the 'volts/cm' switch on 2 V/cm.

Fig 17

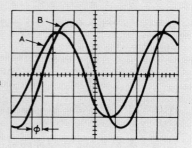

(a) The width of each complete cycle is 5 cm for both waveforms. Hence the periodic time, T, of each waveform is 5 cm $\times$ 100 μs/cm, i.e. 0.5 ms.

Frequency of each waveform, $f = \dfrac{1}{T} = \dfrac{1}{0.5 \times 10^{-3}} = $ **2 kHz**

(b) The peak value of waveform A is 2 cm $\times$ 2 V/cm = 4 V.

Hence the r.m.s. value of waveform A = $\dfrac{4}{\sqrt{2}} = $ **2.83 V**

The peak value of waveform B is 2.5 cm × 2 V/cm = 5 V.

Hence the r.m.s. value of waveform B = $\dfrac{5}{\sqrt{2}}$ = **3.54 V**

(c) Since 5 cm represents 1 cycle, then 5 cm represents 360°, i.e. 1 cm represents $\dfrac{360}{5}$ = 72°. The phase angle ϕ = 0.5 cm = 0.5 cm × 72°/cm = 36°.

Hence waveform A leads waveform B by 36°

Problem 11 In a Wheatstone bridge ABCD, a galvanometer is connected between A and C, and a battery between B and D. A resistor of unknown value is connected between A and B. When the bridge is balanced, the resistance between B and C is 100 Ω, that between C and D is 10 Ω and that between D and A is 400 Ω. Calculate the value of the unknown resistance.

The Wheatstone bridge is shown in *Fig 18* where R_x is the unknown resistance. At balance, equating the products of opposite ratio arms, gives:

$(R_x)(10) = (100)(400)$

$R_x = \dfrac{(100)(400)}{10} = 4000\ \Omega$

Hence unknown resistance, R_x = 4 kΩ

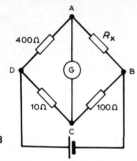

Fig 18

Problem 12 In a d.c. potentiometer balance is obtained at a length of 400 mm when using a standard cell of 1.0186 volts. Determine the e.m.f. of a dry cell if balance is obtained with a length of 650 mm.

$E_1 = 1.0186$ V, $l_1 = 400$ mm, $l_2 = 650$ mm

With reference to *Figure 8*, $\dfrac{E_1}{E_2} = \dfrac{l_1}{l_2}$

from which, $E_2 = E_1 \dfrac{l_2}{l_1} = (1.0186)\left(\dfrac{650}{400}\right) = \textbf{1.655 volts}$

Problem 13 List the errors most likely to occur in measurements made with electrical measuring instruments.

Errors in the limitations of the instrument

(i) The calibration accuracy of an instrument depends on the precision with which it is constructed. Every instrument has a margin of error which is expressed as a percentage of the indication. For example, industrial grade instruments have an accuracy of ± 2% of f.s.d. Thus if a voltmeter has a f.s.d. of 100 V and it indicates 40 V say, then the actual voltage may be anywhere between 40 ± (2% of 100), or 40 ± 2, i.e. between 38 V and 42 V.

   When an instrument is calibrated, it is compared against a standard instrument and a graph is drawn of 'error' against 'meter deflection'. A typical graph is shown

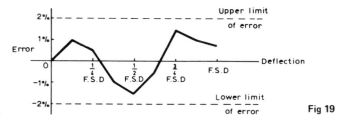

Fig 19

in *Fig 19* where it is seen that the accuracy varies over the scale length. Thus a meter with a ± 2% f.s.d. accuracy would tend to have an accuracy which is much better than ± 2% f.s.d. over much of the range.

Errors by the operator

(ii) It is easy for an operator to misread an instrument reading. With linear scales the values of the sub-divisions are reasonably easy to determine; non-linear scale graduations are more difficult to estimate. Also, scales differ from instrument to instrument and some meters have more than one scale (as with multimeters) and mistakes in reading indications are easily made. When reading a meter scale it should be viewed from an angle perpendicular to the surface of the scale at the location of the pointer; a meter scale should not be viewed 'at an angle'.

Errors due to the instrument disturbing the circuit

(iii) Any instrument connected into a circuit will affect that circuit to some extent. Meters require some power to operate, but provided this power is small compared with the power in the measured circuit, then little error will result. Incorrect positioning of instruments in a circuit can be a source of errors. For example, let a resistance be measured by the voltmeter–ammeter method as shown in *Fig 20*. Assuming 'perfect' instruments, the resistance should be given by the voltmeter reading divided by the ammeter reading

(i.e. $R = \dfrac{V}{I}$). However, in *Fig 20(a)*, $\dfrac{V}{I} = R + r_a$ and in *Fig 20(b)*

the current through the ammeter is that through the resistor plus that through the voltmeter. Hence the voltmeter reading divided by the ammeter reading will not give the true value of the resistance R for either methods of connection.

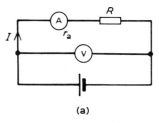

(a)

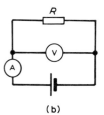

(b)

Fig 20

C. FURTHER PROBLEMS ON ELECTRICAL MEASURING INSTRUMENTS AND MEASUREMENTS

(a) SHORT ANSWER PROBLEMS

1 What is the main difference between an analogue and a digital type measuring instrument?

2 Name the three essential devices for all analogue electrical indicating instruments.

3 Complete the following statements:
(a) An ammeter has a . . . . . . resistance and is connected . . . . . . with the circuit.
(b) A voltmeter has a . . . . . . resistance and is connected . . . . . . with the circuit.

4 State two advantages and two disadvantages of a moving-coil instrument.

5 What effect does the connection of (a) a shunt; (b) a multiplier have on a milliammeter?

6 State two advantages and two disadvantages of a moving-iron instrument.

7 Briefly explain the principle of operation of an ohmmeter.

8 Name a type of ohmmeter used for measuring (a) low resistance values; (b) high resistance values.

9 What is a multimeter?

10 When may a rectifier instrument be used in preference to either the moving-coil or moving-iron instrument?

11 What is the principle of the Wheatstone bridge?

12 How may a d.c. potentiometer be used to measure p.d.s?

13 What is meant by a null method of measurement?

14 Define 'calibration accuracy' as applied to a measuring instrument.

15 State three main areas where errors are most likely to occur in measurements.

16 Name five quantities that a c.r.o. is capable of measuring.

(b) MULTI-CHOICE PROBLEMS (answers on page 172)

1 Which of the following would apply to a moving-coil instrument?
(a) An uneven scale, measuring d.c.; (b) An even scale, measuring a.c.; (c) An uneven scale, measuring a.c.; (d) An even scale, measuring d.c.

2 In *Problem 1*, which would refer to a moving-iron instrument?

3 In *Problem 1*, which would refer to a moving-coil rectifier instrument?

4 Which of the following is needed to extend the range of a milliammeter to read voltages of the order of 100 V?
(a) A parallel high-value resistance; (b) A series high-value resistance; (c) A parallel low-value resistance; (d) A series low-value resistance.

5 *Fig 21* shows a scale of a multi-range ammeter. What is the current indicated when switched to a 25 A scale?
(a) 84 A; (b) 5.6 A; (c) 14 A; (d) 8.4 A.

66

A sinusoidal waveform is displayed on a c.r.o. screen. The peak-to-peak distance is 5 cm and the distance between cycles is 4 cm. The 'variable' switch is on 100 μs/cm and the 'volts/cm' switch is on 10 V/cm. In *Problems 6 to 10*, select the correct answer from the following:
(a) 25 V; (b) 5 V; (c) 0.4 ms; (d) 35.4 V; (e) 4 ms; (f) 50 V; (g) 250 Hz; (h) 2.5 V; (i) 2.5 kHz; (j) 17.7 V.

Fig 21

6 Determine the peak-to-peak voltage.

7 Determine the periodic time of the waveform.

8 Determine the maximum value of voltage.

9 Determine the frequency of the waveform.

10 Determine the r.m.s. value of the waveform.

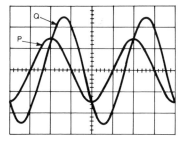

Fig 22

Fig 22 shows double-beam c.r.o. waveform traces. For the quantities stated in *Problems 11 to 17*, select the correct answer from the following:

(a) 30 V; (b) 0.2 s; (c) 50 V; (d) $\dfrac{15}{\sqrt{2}}$ V; (e) 54° leading; (f) $\dfrac{250}{\sqrt{2}}$ V; (g) 15 V;

(h) 100 μs; (i) $\dfrac{50}{\sqrt{2}}$ V; (j) 250 V; (k) 10 kHz; (l) 75 V; (m) 40 μs; (n) $\dfrac{3\pi}{10}$ rads lagging;

(o) $\dfrac{25}{\sqrt{2}}$ V; (p) 5 kHz; (q) $\dfrac{30}{\sqrt{2}}$ V; (r) 25 kHz; (s) $\dfrac{75}{\sqrt{2}}$ V; (t) $\dfrac{3\pi}{10}$ rads leading

11 Amplitude of waveform P.

12 Peak-to-peak value of waveform Q.

13 Periodic time of both waveforms.

14 Frequency of both waveforms.

15 R.M.S. value of waveform P.

16 R.M.S. value of waveform Q.

17 Phase displacement of waveform Q relative to waveform P.

1 A moving-coil instrument gives f.s.d. for a current of 10 mA. Neglecting the resistance of the instrument, calculate the approximate value of series resistance needed to enable the instrument to measure up to (a) 20 V; (b) 100 V; (c) 250 V.

[(a) 2 kΩ; (b) 10 kΩ; (c) 25 kΩ]

2 A meter of resistance 50 Ω has a f.s.d. of 4 mA. Determine the value of shunt resistance required in order that the f.s.d. should be (a) 15 mA; (b) 20 A; (c) 100 A.

[(a) 18.18 Ω; (b) 10.00 mΩ; (c) 2.00 mΩ]

3 A moving-coil instrument having a resistance of 20 Ω, gives a f.s.d. when the current is 5 mA. Calculate the value of the multiplier to be connected in series with the instrument so that it can be used as a voltmeter for measuring p.d.s up to 200 V.

[39.98 kΩ]

4 A moving-coil instrument has a f.s.d. current of 20 mA and a resistance of 25 Ω. Calculate the values of resistance required to enable the instrument to be used (a) as a 0–10 A ammeter, and (b) as a 0–100 V voltmeter. State the mode of resistance connection in each case. [(a) 50.10 mΩ in parallel; (b) 4.975 kΩ in series]

5 A meter has a resistance of 40 Ω and registers a maximum deflection when a current of 15 mA flows. Calculate the value of resistance that converts the movement into (a) an ammeter with a maximum deflection of 50 A; (b) a voltmeter with a range 0–250 V. [(a) 12.00 mΩ in parallel; (b) 16.63 kΩ in series]

6 (a) Describe, with the aid of diagrams, the principle of operation of a moving-iron instrument.
(b) Draw a circuit diagram showing how a moving-coil instrument may be used to measure alternating current.
(c) Discuss the advantages and disadvantages of moving-coil rectifier instruments when compared with moving-iron instruments.

7 (a) Describe, with the aid of a diagram, the principle of the Wheatstone bridge and hence deduce the balance condition giving the unknown resistance in terms of known values of resistance.
(b) In a Wheatstone bridge PQRS, a galvanometer is connected between Q and S and a voltage source between P and R. An unknown resistor R_x is connected between P and Q. When the bridge is balanced, the resistance between Q and R is 200 Ω, that between R and S is 10 Ω and that between S and P is 150 Ω. Calculate the value of R

[3 kΩ]

8 (a) Describe, with the aid of a diagram, how a d.c. potentiometer can be used to measure the e.m.f. of a cell.
(b) Balance is obtained in a d.c. potentiometer at a length of 31.2 cm when using a standard cell of 1.0186 volts. Calculate the e.m.f. of a dry cell if balance is obtained with a length of 46.7 cm. [1.525 V]

9 List the errors most likely to occur in the measurements of electrical quantities.
    A 240 V supply is connected across a load resistance R. Also connected across R is a voltmeter having a f.s.d. of 300 V and a figure or merit (i.e. sensitivity) of 8 kΩ/V. Calculate the power dissipated by the voltmeter and by the load resistance if (a) $R = 100$ Ω; (b) $R = 1$ MΩ. Comment on the results obtained.

[(a) 24 mW, 576 W; (b) 24 mW, 57.6 mW]

10 A 0–1 A ammeter having a resistance of 50 Ω is used to measure the current flowing in a 1 kΩ resistor when the supply voltage is 250 V. Calculate: (a) the approximate value of current (neglecting the ammeter resistance); (b) the actual current in the

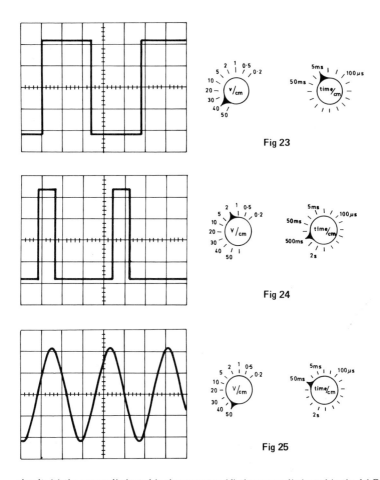

Fig 23

Fig 24

Fig 25

circuit; (c) the power dissipated in the ammeter; (d) the power dissipated in the 1 kΩ
resistor.   [(a) 0.250 A; (b) 0.238 A; (c) 2.832 W; (d) 56.64 W]

11 For the square voltage waveform displayed on a c.r.o. shown in *Fig 23*, find (a)
    its frequency; (b) its peak-to-peak voltage.   [(a) 41.7 Hz; (b) 176 V]

12 For the pulse waveform shown in *Fig 24*, find (a) its frequency; (b) the magnitude
    of the pulse voltage.   [(a) 0.56 Hz; (b) 8.4 V]

13 For the sinusoidal waveform shown in *Fig 25*, determine (a) its frequency; (b) the
    peak-to-peak voltage; (c) the r.m.s. voltage.   [(a) 7.14 Hz; (b) 220 V; (c) 77.78 V]

6 The effects of forces on materials

A. MAIN POINTS CONCERNED WITH THE EFFECTS OF FORCES ON MATERIALS

1 A **force** exerted on a body can cause a change in either the shape or the motion of the body. The unit of force is the **newton, N**.

2 No solid body is perfectly rigid and when forces are applied to it, changes in dimensions occur. Such changes are not always perceptible to the human eye since they are so small. For example, the span of a bridge will sag under the weight of a vehicle and a spanner will bend slightly when tightening a nut. It is important for engineers and designers to appreciate the effects of forces on materials, together with their mechanical properties.

3 The three main types of mechanical force that can act on a body are (i) tensile, (ii) compressive, (iii) shear.

Tensile force

4 Tension is a force which tends to stretch a material, as shown in *Fig 1(a)*. Examples include:
(i) the rope or cable of a crane carrying a load is in tension; (ii) rubber bands, when stretched, are in tension; (iii) a bolt — when a nut is tightened, a bolt is under tension. A tensile force, i.e. one producing tension, increases the length of the material on which it acts.

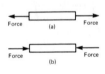

Fig 1

Compressive force

5 Compression is a force which tends to squeeze or crush a material, as shown in *Fig 1(b)*. Examples include:
(i) a pillar supporting a bridge is in compression; (ii) the sole of a shoe is in compression; (iii) the jib of a crane is in compression.
A compressive force, i.e., one producing compression, will decrease the length of the material on which it acts.

Shear force

6 Shear is a force which tends to slide one face of the material over an adjacent face. Examples include:
(i) a rivet holding two plates together is in shear if a tensile force is applied between the plates, as shown in *Fig 2*;

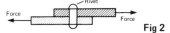

Fig 2

(ii) a guillotine cutting sheet metal, or garden shears each provide a shear force;
(iii) a horizontal beam is subject to shear force;
(iv) transmission joints on cars are subject to
shear forces.

A shear force can cause a material to bend,
slide or twist.
(See *Problem 1*).

7 Forces acting on a material cause a change in dimensions and the material is said to
be in a state of **stress**. Stress is the ratio of the applied force F to cross-sectional
area A of the material. The symbol used for tensile and compressive stress is σ (Greek
letter sigma). The unit of stress is the Pascal Pa, where 1 Pa = 1 N/m². Hence

$$\sigma = \frac{F}{A} \text{ Pa}$$

where F is the force in newtons and A is the cross-sectional area in square metres.
For tensile and compressive forces, the cross-sectional area is that which is at right-
angles to the direction of the force. For a shear force the shear stress is equal to F/A,
where the cross-sectional area A is that which is parallel to the direction of the
force. The symbol used for shear stress is the Greek letter tau, τ.

8 The fractional change in a dimension of a material produced by a force is called the
strain. For a tensile or compressive force, strain is the ratio of the change of length
to the original length. The symbol used for strain is ϵ (Greek epsilon). For a material
of length l metres which changes in length by an amount x metres when subjected to
stress,

$$\epsilon = \frac{x}{l}$$

Strain is dimensionless and is often expressed as a percentage,

i.e. $$\text{percentage strain} = \frac{x}{l} \times 100$$

For a shear force, strain is denoted
by the symbol γ (Greek letter gamma)
and, with reference to *Fig 3*, is
given by: $$\gamma = \frac{x}{l}$$

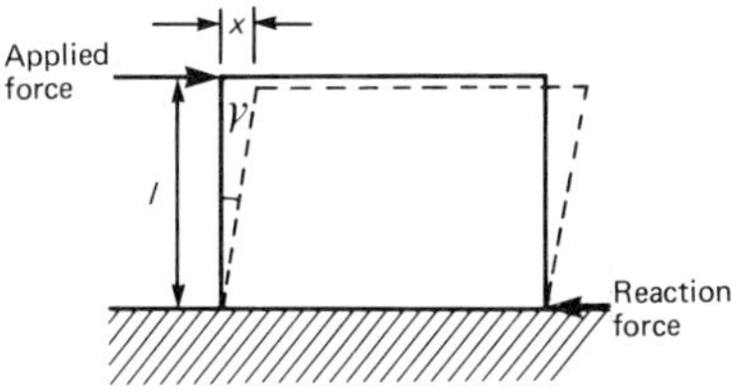

Fig 3

(see *Problems 2 to 9*)

9 **Elasticity** is the ability of a material to return to its original shape and size on the
removal of external forces. If it does not return to the original shape, it is said to be
plastic. Within certain load limits mild steel, copper and rubber are examples of
elastic materials; lead and plasticine are examples of plastic materials.

10 If a tensile force applied to a uniform bar of mild steel is gradually increased
and the corresponding extension of the bar is measured, then provided the
applied force is not too large, a graph depicting these results is likely to be

as shown in *Fig 4*. Since the graph is a straight line, **extension is directly proportional to the applied force.**

11 If the applied force is large, it is found that the material no longer returns to its original length when the force is removed. The material is then said to have passed its **elastic limit** and the resulting graph of force/extension is no longer a straight line (see para. 15). Stress $\sigma = F/A$, from para. 7, and since, for a particular bar, A can be considered as constant, then $F \propto \sigma$. Strain $\epsilon = x/l$, from para. 8, and since for a particular bar l is constant, then $x \propto \epsilon$ Hence for stress applied to a material below the elastic limit a graph of stress/strain will be as shown in *Fig 5*, and is a similar shape to the force/extension graph of *Fig 4*.

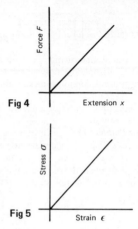

Fig 4

Fig 5

12 **Hooke's law** states:

'Within the elastic limit, the extension of a material is proportional to the applied force.'

It follows, from para. 11, that:

'Within the elastic limit of a material, the strain produced is directly proportional to the stress producing it.'

13 Within the elastic limit, stress $\propto$ strain,

hence stress = (a constant) $\times$ strain

This constant of proportionality is called **Young's Modulus of Elasticity** and is given the symbol E. The value of E may be determined from the gradient of the straight line portion of the stress/strain graph. The dimensions of E are pascals (the same as for stress, since strain is dimensionless).

$$E = \frac{\sigma}{\epsilon} \text{ Pa}$$

Some typical values for Young's modulus of elasticity, E include:
Aluminium 70 G Pa (i.e. 70×10^9 Pa), brass 100 G Pa, copper 110 G Pa, diamond 1200 G Pa, mild steel 210 G Pa, lead 18 G Pa, tungsten 410 G Pa, cast iron 110 G Pa, zinc 110 G Pa.

14 A material having a large value of Young's modulus is said to have a high value of stiffness, where stiffness is defined as:

Stiffness $= \dfrac{\textbf{force } F}{\textbf{extension } x}$

For example, mild steel is much stiffer than lead.

Since $E = \dfrac{\sigma}{\epsilon}$ and $\sigma = \dfrac{F}{A}$ and $\epsilon = \dfrac{x}{l}$

then $E = \dfrac{F/A}{x/l}$, i.e. $E = \dfrac{Fl}{Ax} = \left(\dfrac{F}{x}\right)\left(\dfrac{l}{A}\right)$

i.e. $E = $ **(stiffness)** $\times \left(\dfrac{l}{A}\right)$

72

Stiffness $\left(=\dfrac{F}{x}\right)$ is also the gradient of the force/extension graph, hence

$E =$ (gradient of force/extension graph)$\left(\dfrac{l}{A}\right)$

Since l and A for a particular specimen are constant, the greater Young's modulus the greater the stiffness.

(See *Problems 10 to 16*).

15 A **tensile test** is one in which a force is applied to a specimen of a material in increments and the corresponding extension of the specimen noted. The process may be continued until the specimen breaks into two parts and this is called testing to destruction. The testing is usually carried out using a universal testing machine which can apply either tensile or compressive forces to a specimen in small, accurately-measured steps. **British Standard 18** gives the standard procedure for such a test. Test specimens of a material are made to standard shapes and sizes and two typical test pieces are shown in *Fig 6*. The results of a tensile test may be plotted on a load/extension graph and a typical graph for a mild steel specimen is shown in *Fig 7*.

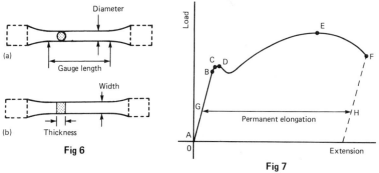

Fig 6

Fig 7

(i) Between A and B is the region in which Hooke's law applies and stress is directly proportional to strain. The gradient of AB is used when determining Young's modulus of elasticity (see para. 14).

(ii) Point B is the **limit of proportionality** and is the point at which stress is no longer proportional to strain when a further load is applied.

(iii) Point C is the **elastic limit** and a specimen loaded to this point will effectively return to its original length when the load is removed, i.e. there is negligible permanent extension.

(iv) Point D is called the **yield point** and at this point there is a sudden extension with no increase in load.

(v) Between D and E extension takes place over the whole gauge length of the specimen.

(vi) Point E gives the maximum load which can be applied to the specimen and is used to determine the ultimate tensile strength (U.T.S.) of the specimen.

U.T.S. $= \dfrac{\text{maximum load}}{\text{original cross-sectional area}}$

(vii) Between points E and F the cross-sectional area of the specimen decreases, usually about half way between the ends, and a **waist** or **neck** is formed before fracture. The value of stress at F is greater than at E since although the load on the specimen is decreasing as the extension increases, the cross-sectional area is also reducing.

(viii) At point F the specimen fractures.

(ix) Distance GH is called the **permanent elongation** and

$$\text{Percentage elongation} = \frac{\text{increase in length during test to destruction}}{\text{original length}} \times 100\%$$

(See *Problems 17 to 19*)

16 (i) **Ductility** is the ability of a material to be permanently stretched (i.e. drawn out to a small cross section by a tensile force). For ductile materials such as mild steel, copper and gold, large extensions can result before fracture occurs. Ductile materials usually have a percentage elongation value of about 15% or more.

(ii) **Brittleness** is a lack of ductility. Brittle materials such as cast iron have virtually no plastic stage. The elastic stage is followed by immediate fracture. Other examples of brittle materials include glass, concrete, brick and ceramics. Little or no 'waist' occurs before fracture in a brittle material undergoing a tensile test and there is no noticeable yield point.

(See *Problem 20*)

B. WORKED PROBLEMS ON THE EFFECTS OF FORCES ON MATERIALS

Problem 1 Fig 8(a)
represents a crane and
Fig 8(b) a transmission joint.
State the types of forces acting,
labelled A to F.

Fig 8

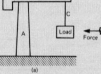
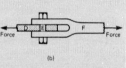

(a) For the crane, A, a supporting member, is in **compression,** B, a horizontal beam, is in **shear,** and C, a rope, is in **tension.**

(b) For the transmission joint, parts D and F are in **tension,** and E, the rivet or bolt, is in **shear.**

Problem 2 A rectangular bar having a cross-sectional area of 75 mm² has a tensile force of 15 kN applied to it. Determine the stress in the bar.

Cross-sectional area $A = 75 \text{ mm}^2 = 75 \times 10^{-6} \text{ m}^2$; Force $F = 15 \text{ kN} = 15 \times 10^3 \text{ N}$

Stress in bar, $\sigma = \dfrac{F}{A} = \dfrac{15 \times 10^3 \text{ N}}{75 \times 10^{-6} \text{ m}} = 0.2 \times 10^9 \text{ Pa} = \textbf{200 MPa}$

Problem 3 A circular wire has a tensile force of 60.0 N applied to it and this force produces a stress of 3.06 MPa in the wire. Determine the diameter of the wire.

Force $F = 60.0 \text{ N}$; Stress $\sigma = 3.06 \text{ MPa} = 3.06 \times 10^6 \text{ Pa}$

Since $\sigma = \dfrac{F}{A}$, then area $A = \dfrac{F}{\sigma} = \dfrac{60.0 \text{ N}}{3.06 \times 10^6 \text{ Pa}} = 19.61 \times 10^{-6} \text{ m}^2$

$$= 19.61 \text{ mm}^2$$

Cross-sectional area $A = \dfrac{\pi d^2}{4}$. Hence $19.61 = \dfrac{\pi d^2}{4}$

from which, $d^2 = \dfrac{4 \times 19.61}{\pi}$ and $d = \sqrt{\left(\dfrac{4 \times 19.61}{\pi}\right)}$

i.e. **diameter of wire = 5.0 mm**

Problem 4 A bar 1.60 m long contracts by 0.1 mm when a compressive load is applied to it. Determine the strain and the percentage strain.

Strain $\epsilon = \dfrac{\text{contraction}}{\text{original length}} = \dfrac{0.1 \text{ mm}}{1.60 \times 10^3 \text{ mm}} = \dfrac{0.1}{1600} = \mathbf{0.000\ 062\ 5}$

Percentage strain $= 0.000\ 062\ 5 \times 100 = \mathbf{0.006\ 25\%}$

Problem 5 A wire of length 2.50 m has a percentage strain of 0.012% when loaded with a tensile force. Determine the extension in the wire.

Original length of wire $= 2.50 \text{ m} = 2500 \text{ mm}$

Strain $= \dfrac{0.012}{100} = 0.000\ 12$

Strain $\epsilon = \dfrac{\text{extension } x}{\text{original length } l}$, hence extension $x = \epsilon l = (0.000\ 12)(2500)$

$$= \mathbf{0.3 \text{ mm}}$$

Problem 6 (a) A rectangular metal bar has a width of 10 mm and can support a maximum compressive stress of 20 MPa. Determine the minimum breadth of the bar when loaded with a force of 3 kN.
(b) If the bar in (a) is 2 m long and decreases in length by 0.25 mm when the force is applied, determine the strain and the percentage strain.

(a) Since stress $\sigma = \dfrac{\text{force } F}{\text{area } A}$, then area $A = \dfrac{F}{\sigma} = \dfrac{3000 \text{ N}}{20 \times 10^6 \text{ Pa}} = 150 \times 10^{-6} \text{ m}^2$

$$= 150 \text{ mm}^2$$

Cross-sectional area $=$ width $\times$ breadth, hence breadth $= \dfrac{\text{area}}{\text{width}} = \dfrac{150}{10} = \mathbf{15 \text{ mm}}$

(b) Strain $\epsilon = \dfrac{\text{contraction}}{\text{original length}} = \dfrac{0.25}{2000} = \mathbf{0.000\ 125}$

Percentage strain $= 0.000\ 125 \times 100 = \mathbf{0.0125\%}$

Problem 7 A pipe has an outside diameter of 25 mm, an inside diameter of 15 mm and length 0.40 m and it supports a compressive load of 40 kN. The pipe shortens by 0.5 mm when the load is applied. Determine (a) the compressive stress; (b) the compressive strain in the pipe when supporting this load.

Compressive force $F = 40 \text{ kN} = 40\ 000 \text{ N}$

Cross-sectional area of pipe $A = \dfrac{\pi D^2}{4} - \dfrac{\pi d^2}{4}$, where $D =$ outside diameter $= 25$ mm and $d =$ inside diameter $= 15$ mm.

Hence $A = \dfrac{\pi}{4}(25^2 - 15^2) \text{ mm}^2 = \dfrac{\pi}{4}(25^2 - 15^2) \times 10^{-6} \text{ m}^2 = 3.142 \times 10^{-4} \text{ m}^2$

(a) Compressive stress $\sigma = \dfrac{F}{A} = \dfrac{40\ 000 \text{ N}}{3.142 \times 10^{-4} \text{ m}^2} = 12.73 \times 10^7 \text{ Pa} = \mathbf{127.3 \text{ MPa}}$

(b) Contraction of pipe when loaded, $x = 0.5$ mm $= 0.0005$ m
Original length of pipe, $l = 0.4$ m

Hence compressive strain $\epsilon = \dfrac{x}{l} = \dfrac{0.0005}{0.4} = \mathbf{0.001\ 25}$ (or 0.125%)

Problem 8 A circular hole of diameter 50mm is to be punched out of a 2mm thick metal plate. The shear stress needed to cause fracture is 500MPa. Determine (a) the minimum force to be applied to the punch; (b) the compressive stress in the punch at this value.

(a) The area of metal to be sheared, A = perimeter of hole × thickness of plate.
Perimeter of hole $= \pi d = \pi(0.050) = 0.1571$ m
Hence shear area $A = 0.1571 \times 0.002 = 314.2 \times 10^{-6}$ m^2

Since shear stress $= \dfrac{\text{force}}{\text{area}}$, shear force = shear stress × area

$= (500 \times 10^6 \times 314.2 \times 10^{-6})$ N $= 157.1$ kN, which is the minimum force to be applied to the punch.

(b) Area of punch $= \dfrac{\pi d^2}{4} = \dfrac{\pi(0.050)^2}{4} = 0.001\ 963$ m^2

Compressive stress $= \dfrac{\text{force}}{\text{area}} = \dfrac{157.1 \times 10^3 \text{ N}}{0.001963 \text{ m}^2} = 8.003 \times 10^7$ Pa

$= \mathbf{80.03\ MPa}$, which is the compressive stress in the punch.

Problem 9 A rectangular block of plastic material 500 mm long by 20 mm wide by 300 mm high has its lower face glued to a bench and a force of 200 N is applied to the upper face and in line with it. The upper face moves 15 mm relative to the lower face. Determine (a) the shear stress, and (b) the shear strain in the upper face, assuming the deformation is uniform.

(a) Shear stress

$$\tau = \frac{\text{force}}{\text{area parallel to the force}}$$

Area of any face parallel to the
force = 500 mm × 20 mm
$= (0.5 \times 0.02)$ m$^2 = 0.01$ m^2.

Hence shear stress $\tau = \dfrac{200 \text{ N}}{0.01 \text{ m}^2}$

$= \mathbf{20\ 000\ Pa}$ or $\mathbf{20\ kPa}$

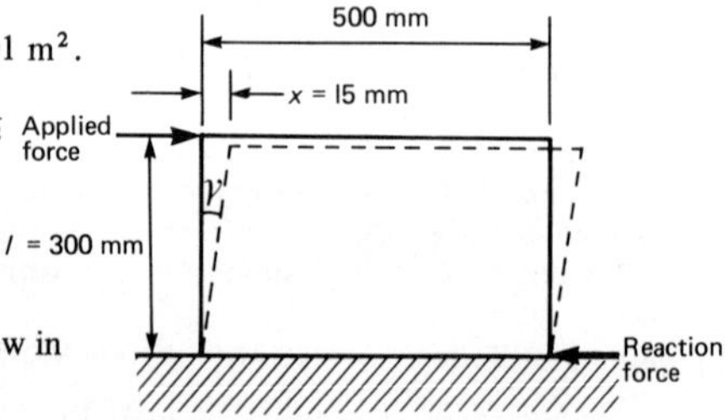

(b) Shear strain $\gamma = \dfrac{x}{l}$ (see side view in *Fig 9*) $= \dfrac{15}{300} = \mathbf{0.05}$ (or 5%).

Fig 9

Problem 10 A wire is stretched 2 mm by a force of 250 N. Determine the force that would stretch the wire 5 mm, assuming that the elastic limit is not exceeded.

Hooke's law states that extension x is proportional to force F, provided that the elastic limit is not exceeded, i.e., $x \propto F$ or $x = kF$ where k is a constant.

When $x = 2$ mm, $F = 250$ N, thus $2 = k(250)$,

from which, constant $k = \dfrac{2}{250} = \dfrac{1}{125}$

When $x = 5$ mm, then $5 = kF$, i.e. $5 = \left(\dfrac{1}{125}\right)F$

from which, force $F = 5(125) = 625$ N

Thus to stretch the wire 5 mm a force of 625 N is required

Problem 11 A force of 10 kN applied to a component produces an extension of 0.1 mm. Determine (a) the force needed to produce an extension of 0.12 mm, and (b) the extension when the applied force is 6 kN, assuming in each case that the elastic limit is not exceeded.

From Hooke's law, extension x is proportional to force F within the elastic limit, i.e., $x \propto F$ or $x = kF$, where k is a constant. If a force of 10 kN produces an extension of 0.1 mm, then $0.1 = k(10)$

from which, constant $k = \dfrac{0.1}{10} = 0.01$

(a) When extension $x = 0.12$ mm, then $0.12 = k(F)$, i.e. $0.12 = 0.01\ F$

from which, force $F = \dfrac{0.12}{0.01} = \textbf{12 kN}$

(b) When force $F = 6$ kN, then **extension** $x = k(6) = (0.01)(6) = \textbf{0.06 mm}$

Problem 12 A copper rod of diameter 20 mm and length 2.0 m has a tensile force of 5 kN applied to it. Determine (a) the stress in the rod; (b) by how much the rod extends when the load is applied. Take the modulus of elasticity for copper as 96 GPa.

(a) Force $F = 5$ kN $= 5000$ N

Cross-sectional area $A = \dfrac{\pi d^2}{4} = \dfrac{\pi(0.020)^2}{4} = 0.000\ 314$ m^2

Stress $\sigma = \dfrac{F}{A} = \dfrac{5000\ \text{N}}{0.000\ 314\ \text{m}^2} = 15.92 \times 10^6$ Pa $= \textbf{15.92 MPa}$

(b) Since $E = \dfrac{\sigma}{\epsilon}$ then strain $\epsilon = \dfrac{\sigma}{E} = \dfrac{15.92 \times 10^6\ \text{Pa}}{96 \times 10^9\ \text{Pa}} = 0.000\ 166$

Strain $\epsilon = \dfrac{x}{l}$, hence extension, $x = \epsilon l = (0.000\ 166)(2.0) = 0.000\ 332$ m

i.e. extension of rod is 0.332 mm

Problem 13 A bar of thickness 15 mm and having a rectangular cross-section carries a load of 120 kN. Determine the minimum width of the bar to limit the maximum stress to 200 MPa. The bar which is 1.0 m long extends by 2.5 mm when carrying a load of 120 kN. Determine the modulus of elasticity of the material of the bar.

Force, $F = 120$ kN $= 120\ 000$ N

Cross-sectional area $A = (15x)10^{-6}$ m^2, where x is the width of the rectangular bar in millimetres.

Stress $\sigma = \dfrac{F}{A}$, from which $A = \dfrac{F}{\sigma} = \dfrac{120\ 000\ \text{N}}{200 \times 10^6\ \text{Pa}} = 6 \times 10^{-4}$ m$^2 = 6 \times 10^2$ mm^2

$= 600$ mm^2

Hence $600 = 15x$, from which, width of bar $x = \frac{600}{15} = \textbf{40 mm}$

Extension of bar $= 2.5$ mm $= 0.0025$ m

Strain $\epsilon = \frac{x}{l} = \frac{0.0025}{1.0} = 0.0025$

Modulus of elasticity $E = \frac{\text{stress}}{\text{strain}} = \frac{200 \times 10^6}{0.0025} = 80 \times 10^9 = \textbf{80 GPa}$

Problem 14 An aluminium rod has a length of 200 mm and a diameter of 10 mm. When subjected to a compressive force the length of the rod is 199.6 mm. Determine (a) the stress in the rod when loaded, and (b) the magnitude of the force. Take the modulus of elasticity for aluminium as 70 GPa.

(a) Original length of rod, $l = 200$ mm; final length of rod $= 199.6$ mm
Hence contraction, $x = 0.4$ mm.

Thus strain $\epsilon = \frac{x}{l} = \frac{0.4}{200} = 0.002$

Modulus of elasticity, $E = \frac{\text{stress } \sigma}{\text{strain } \epsilon}$, hence stress $\sigma = E\epsilon = 70 \times 10^9 \times 0.002$
$$= 140 \times 10^6 \text{ Pa} = \textbf{140 MPa}$$

(b) Since stress $\sigma = \frac{\text{force } F}{\text{area } A}$, then force $F = \sigma A$

Cross-sectional area, $A = \frac{\pi d^2}{4} = \frac{\pi (0.010)^2}{4} = 7.854 \times 10^{-5}$ m²

Hence compressive force $F = \sigma A = 140 \times 10^6 \times 7.854 \times 10^{-5} = \textbf{11.0 kN}$

Problem 15 A brass tube has an internal diameter of 120 mm and an outside diameter of 150 mm and is used to support a load of 5 kN. The tube is 500 mm long before the load is applied. Determine by how much the tube contracts when loaded, taking the modulus of elasticity for brass as 90 GPa.

Force in tube $F = 5$ kN $= 5000$ N

Cross-sectional area of tube, $A = \frac{\pi}{4}(D^2 - d^2) = \frac{\pi}{4}(0.150^2 - 0.120^2) = 0.006\,362$ m²

Stress in tube, $\sigma = \frac{F}{A} = \frac{5000 \text{ N}}{0.006\,362 \text{ m}^2} = 0.7859 \times 10^6$ Pa

Since the modulus of elasticity, $E = \frac{\text{stress } \sigma}{\text{strain } \epsilon}$, then strain $\epsilon = \frac{\sigma}{E}$

$$= \frac{0.7859 \times 10^6 \text{ Pa}}{90 \times 10^9 \text{ Pa}} = 8.732 \times 10^{-6}.$$

Strain $\epsilon = \frac{\text{contraction } x}{\text{original length } l}$, thus contraction $x = \epsilon l = 8.732 \times 10^{-6} \times 0.500$
$$= 4.37 \times 10^{-6} \text{ m}$$

Thus, when loaded, the tube contracts by 4.37 μm

Problem 16 In an experiment to determine the modulus of elasticity of a sample of mild steel, a wire is loaded and the corresponding extension noted. The results of the experiment are as shown.

Load (N)	0	40	110	160	200	250	290	340
Extension (mm)	0	1.2	3.3	4.8	6.0	7.5	10.0	16.2

Draw the load/extension graph. The mean diameter of the wire is 1.3 mm and its length is 8.0 m. Determine the modulus of elasticity of the sample, and the stress at the limit of proportionality.

A graph of load/extension is shown in *Fig 10*.

From para. 13, $E = \dfrac{\sigma}{\epsilon} = \dfrac{F/A}{x/l} = \left(\dfrac{F}{x}\right)\left(\dfrac{l}{A}\right)$.

$\dfrac{F}{x}$ is the gradient of the straight line part of the load/extension graph.

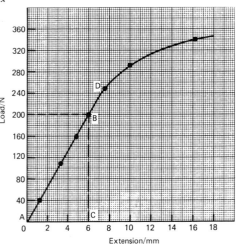

Fig 10

Gradient $\dfrac{F}{x} = \dfrac{BC}{AC} = \dfrac{200\ \text{N}}{6 \times 10^{-3}\ \text{m}} = 33.33 \times 10^3\ \text{N/m}$

Modulus of elasticity $E = (\text{gradient of graph}) \left(\dfrac{l}{A}\right)$

Length of specimen, $l = 8.0$ m

Cross-sectional area $A = \dfrac{\pi d^2}{4} = \dfrac{\pi (0.0013)^2}{4} = 1.327 \times 10^{-6}$

Hence modulus of elasticity $= (33.33 \times 10^3)\left(\dfrac{8.0}{1.327 \times 10^{-6}}\right) = 201$ GPa

The limit of proportionality is at point D in *Fig 10* where the graph no longer follows a straight line. This point corresponds to a load of 250 N as shown.

Stress at limit of proportionality $= \dfrac{\text{force}}{\text{area}} = \dfrac{250}{1.327 \times 10^{-6}} = 188.4 \times 10^6$ Pa

$= \mathbf{188.4\ MPa}$

Problem 17 A tensile test is carried out on a mild steel specimen of gauge length 40 mm and cross-sectional area 100 mm². The results obtained for the specimen up to its yield point are given below.

Load (kN)	0	8	19	29	36
Extension (mm)	0	0.015	0.038	0.060	0.072

The maximum load carried by the specimen is 50 kN and its length after fracture is 52 mm. Determine (a) the modulus of elasticity; (b) the ultimate tensile strength; (c) the percentage elongation of the mild steel.

The load/extension graph is shown in *Fig 11*.

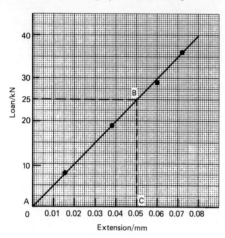

Fig 11

(a) Gradient of straight line $= \dfrac{BC}{AB} = \dfrac{25\,000}{0.05 \times 10^{-3}} = 500 \times 10^6$ N/m

Young's modulus of elasticity $=$ (gradient of graph)$\left(\dfrac{l}{A}\right)$

$l = 40$ mm (gauge length) $= 0.040$ m; Area, $A = 100$ mm² $= 100 \times 10^{-6}$ m²

Young's modulus of elasticity $= (500 \times 10^6)\left(\dfrac{0.040}{100 \times 10^{-6}}\right) = 200 \times 10^9$ Pa
$= \mathbf{200\ GPa}$

(b) Ultimate tensile strength $= \dfrac{\text{maximum load}}{\text{original cross-sectional area}}$

$= \dfrac{50\,000 \text{ N}}{100 \times 10^{-6} \text{ m}^2} = 500 \times 10^6$ Pa $= \mathbf{500\ MPa}$

(c) Percentage elongation $= \dfrac{\text{increase in length}}{\text{original length}} \times 100 = \dfrac{52-40}{40} \times 100 = \dfrac{12}{40} \times 100$
$= \mathbf{30\%}$

Problem 18 The results of a tensile test are:
Diameter of specimen 15 mm; gauge length 40 mm; load at limit of proportionality 85 kN; extension at limit of proportionality 0.075 mm; maximum load 120 kN; final length at point of fracture 55 mm.

Determine (a) Young's modulus of elasticity; (b) the ultimate tensile strength; (c) the stress at the limit of proportionality; (d) the percentage elongation.

(a) Young's modulus of elasticity $E = \dfrac{\text{stress}}{\text{strain}} = \dfrac{F/A}{x/l} = \dfrac{Fl}{Ax}$,

where the load at the limit of proportionality, $F = 85$ kN $= 85\ 000$ N,
l = gauge length = 40 mm = 0.040 m,

A = cross sectional area = $\dfrac{\pi d^2}{4} = \dfrac{\pi (0.015)^2}{4} = 0.000\ 176\ 7$ m^2

x = extension = 0.075 mm = 0.000 075 m

Hence Young's modulus of elasticity $E = \dfrac{Fl}{Ax} = \dfrac{(85\ 000)(0.040)}{(0.000\ 176\ 7)(0.000\ 075)}$

$$= 256.6 \times 10^9 \text{ Pa} = \textbf{256.6 GPa}$$

(b) Ultimate tensile strength $= \dfrac{\text{maximum load}}{\text{original cross-sectional area}}$

$$= \dfrac{120\ 000}{0.000\ 176\ 7} = 679 \times 10^6 \text{ Pa} = \textbf{679 MPa}$$

(c) Stress at limit of proportionality $= \dfrac{\text{load at limit of proportionality}}{\text{cross-sectional area}}$

$$= \dfrac{85\ 000}{0.000\ 176\ 7} = 481.0 \times 10^6 \text{ Pa} = \textbf{481.0 MPa}$$

(d) Percentage elongation $= \dfrac{\text{increase in length}}{\text{original length}} \times 100 = \dfrac{(55-40) \text{ mm}}{40 \text{ mm}} \times 100$

$$= \textbf{37.5\%}$$

Problem 19 A rectangular zinc specimen is subjected to a tensile test and the data from the test is shown below.
Width of specimen 40 mm; breadth of specimen 2.5 mm; gauge length 120 mm.

Load (kN)	10	17	25	30	35	37.5	38.5	37	34	32
Extension (mm)	0.15	0.25	0.35	0.55	1.0	1.50	2.50	3.50	4.50	5.0

Fracture occurs when the extension is 5.0 mm and the maximum load recorded is 38.5 kN.
Plot the load/extension graph and hence determine (a) the stress at the limit of proportionality; (b) Young's modulus of elasticity; (c) the ultimate tensile strength; (d) the percentage elongation; (e) the stress at a strain of 0.01; (f) the extension at a stress of 200 MPa.

A load/extension graph is shown in *Fig 12*.

(a) The limit of proportionality occurs at point P on the graph, where the initial gradient of the graph starts to change. This point has a load value of 26.5 kN.
Cross-sectional area of specimen = 40 mm$\times$2.5 mm = 100 mm^2 = 100×10^{-6} m^2

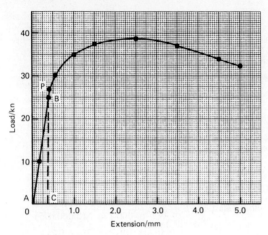

Fig 12

Stress at the limit of proportionality, $\sigma = \dfrac{\text{force}}{\text{area}} = \dfrac{26.5 \times 10^3 \text{ N}}{100 \times 10^{-6} \text{ m}^2}$

$$= 265 \times 10^6 \text{ Pa} = \textbf{265 MPa}$$

(b) Gradient of straight line portion of graph $= \dfrac{\text{BC}}{\text{AC}} = \dfrac{25\ 000 \text{ N}}{0.35 \times 10^{-3} \text{ m}}$

$$= 71.43 \times 10^6 \text{ N/m}$$

Young's modulus of elasticity $= (\text{gradient of graph}) \left(\dfrac{l}{A}\right)$

$$= (71.43 \times 10^6) \left(\dfrac{120 \times 10^{-3}}{100 \times 10^{-6}}\right)$$

$$= 85.72 \times 10^9 \text{ Pa} = \textbf{85.72 GPa}$$

(c) Ultimate tensile strength $= \dfrac{\text{maximum load}}{\text{original cross-sectional area}}$

$$= \dfrac{38.5 \times 10^3 \text{ N}}{100 \times 10^{-6} \text{ m}^2} = 385 \times 10^6 \text{ Pa} = \textbf{385 MPa}$$

(d) Percentage elongation $= \dfrac{\text{extension at fracture point}}{\text{original length}} \times 100$

$$= \dfrac{5.0 \text{ mm}}{120 \text{ mm}} \times 100 = \textbf{4.17\%}$$

(e) Strain $\epsilon = \dfrac{\text{extension } x}{\text{original length } l}$, from which, extension $x = \epsilon l = 0.01 \times 120 = 1.20$ mm

From the graph, the load corresponding to an extension of 1.20 mm is 36 kN.

Stress at a strain of 0.01, $\sigma = \dfrac{\text{force}}{\text{area}} = \dfrac{36\ 000 \text{ N}}{100 \times 10^{-6} \text{ m}^2} = 360 \times 10^6 \text{ Pa}$

$$= \textbf{360 MPa}$$

(f) When the stress is 200 MPa, then force $= \text{area} \times \text{stress}$
$$= (100 \times 10^{-6})(200 \times 10^6) = 20 \text{ kN}$$
From the graph, the corresponding extension is **0.30 mm**

Problem 20 Sketch typical load extension curves for (a) an elastic non-metallic material, (b) a brittle material and (c) a ductile material. Give a typical example of each type of material.

(a) A typical load extension curve for an elastic non-metallic material is shown in *Fig 13(a)*, and an example of such a material is **polythene**.

Fig 13

(b) A typical load extension curve for a brittle material is shown in *Fig 13(b)*, and an example of such a material is **cast iron**.

(c) A typical load extension curve for a ductile material is shown in *Fig 13(c)*, and an example of such a material is **mild steel**.

C. FURTHER PROBLEMS ON THE EFFECTS OF FORCES ON MATERIALS

(a) SHORT ANSWER PROBLEMS

1 What is a tensile force? Name two practical examples of such a force.

2 What is a compressive force? Name two practical examples of such a force.

3 Define a shear force and name two practical examples of such a force.

4 Define elasticity.

5 State Hooke's law.

6 What is the difference between a ductile and a brittle material?

7 Define stress. What is the symbol used for (a) a tensile stress; (b) a shear stress?

8 Strain is the ratio —————— .

9 The ratio $\frac{\text{stress}}{\text{strain}}$ is called . . . .

10 State the units of (a) stress; (b) strain; (c) Young's modulus of elasticity.

11 Stiffness is the ratio —————— .

12 What is a tensile test?

13 Which British Standard gives the standard procedure for a tensile test?

14 With reference to a load/extension graph for mild steel state the meaning of (a) the limit of proportionality; (b) the elastic limit; (c) the yield point; (d) the percentage elongation.

15 Ultimate tensile strength is the ratio —————— .

16 Define (a) ductility; (b) brittleness.

1   A wire is stretched 3 mm by a force of 150 N. Assuming the elastic limit is not exceeded, the force that will stretch the wire 5 mm is: (a) 150 N; (b) 250 N; (c) 90 N

2   For the wire in *Problem 1*, the extension when the applied force is 450 N is: (a) 1 mm; (b) 3 mm; (c) 9 mm.

3   Due to the forces acting a horizontal beam is in: (a) tension; (b) compression; (c) shear.

4   Due to forces acting, a pillar supporting a bridge is in: (a) tension; (b) compression; (c) shear.

5   Which of the following statements is false?
(a) Elasticity is the ability of a material to return to its original dimensions after deformation by a load.
(b) Plasticity is the ability of a material to retain any deformation produced in it by a load.
(c) Ductility is the ability to be permanently stretched without fracturing.
(d) Brittleness is a lack of ductility and a brittle material has a long plastic stage.

6   A circular rod of cross-sectional area 100 mm² has a tensile force of 100 kN applied to it. The stress in the rod is:
(a) 1 MPa; (b) 1 GPa; (c) 1 kPa; (d) 100 MPa.

7   A metal bar 5.0 m long extends by 0.05 mm when a tensile load is applied to it. The percentage strain is:
(a) 0.1; (b) 0.01; (c) 0.001; (d) 0.0001.

An aluminium rod of length 1.0 m and cross-sectional area 500 mm² is used to support a load of 5 kN which causes the rod to contract by 100 μm. For *Problems 8 to 10*, select the correct answer from the following list:
(a) 100 MPa; (b) 0.001; (c) 10 kPa; (d) 100 GPa; (e) 0.01; (f) 10 MPa; (g) 10 GPa; (h) 0.0001; (i) 10 Pa.

8   The stress in the rod.
9   The strain in the rod.
10 Young's modulus of elasticity.

(c) CONVENTIONAL PROBLEMS

1   A rectangular bar having a cross-sectional area of 80 mm² has a tensile force of 20 kN applied to it. Determine the stress in the bar.   [250 MPa]

2   A circular cable has a tensile force of 1 kN applied to it and the force produces a stress of 7.8 MPa in the cable. Calculate the diameter of the cable.   [12.78 mm]

3   A square-sectioned support of side 12 mm is loaded with a compressive force of 10 kN. Determine the compressive stress in the support.   [69.44 MPa]

4   A bolt having a diameter of 5 mm is loaded so that the shear stress in it is 120 MPa. Determine the value of the shear force on the bolt.   [2.356 kN]

5   A split pin requires a force of 400 N to shear it. The maximum shear stress before shear occurs is 120 MPa. Determine the minimum diameter of the pin.   [2.06 mm]

6   A wire of length 4.5 m has a percentage strain of 0.050% when loaded with a tensile force. Determine the extension in the wire.   [2.25 mm]

7 A tube of outside diameter 60 mm and inside diameter 40 mm is subjected to a load of 60 kN. Determine the stress in the tube.   [38.2 MPa]

8 A metal bar 2.5 m long extends by 0.05 mm when a tensile load is applied to it. Determine (a) the strain; (b) the percentage strain.   [(a) 0.000 02; (b) 0.002%]

9 Explain, using appropriate practical examples, the difference between tensile, compressive and shear forces.

10 (a) State Hooke's law.
   (b) A wire is stretched 1.5 mm by a force of 300 N. Determine the force that would stretch the wire 4 mm, assuming the elastic limit of the wire is not exceeded.
                                                                                    [(b) 800 N]

11 A rubber band extends 50 mm when a force of 300 N is applied to it. Assuming the band is within the elastic limit, determine the extension produced by a force of 60 N.   [10 mm]

12 A force of 25 kN applied to a piece of steel produces an extension of 2 mm. Assuming the elastic limit is not exceeded, determine (a) the force required to produce an extension of 3.5 mm; (b) the extension when the applied force is 15 kN.
                                                                      [(a) 43.75 N; (b) 1.2 mm]

13 A coil spring 300 mm long when unloaded, extends to a length of 500 mm when a load of 40 N is applied. Determine the length of the spring when a load of 15 kN is applied.   [375 mm]

14 Sketch on the same axes typical load extension graphs for (a) a strong, ductile material; (b) a brittle material.

15 Define (a) tensile stress; (b) shear stress; (c) strain; (d) shear strain; (e) Young's modulus of elasticity.

16 A circular bar is 2.5 m long and has a diameter of 60 mm. When subjected to a compressive load of 30 kN it shortens by 0.20 mm. Determine Young's modulus of elasticity for the material of the bar.   [132.6 GPa]

17 A bar of thickness 20 mm and having a rectangular cross-section carries a load of 82.5 kN. Determine (a) the minimum width of the bar to limit the maximum stress to 150 MPa; (b) the modulus of elasticity of the material of the bar if the 150 mm long bar extends by 0.8 mm when carrying a load of 200 kN.
                                                                      [(a) 27.5 mm; (b) 68.2 GPa]

18 A metal rod of cross-sectional area 100 mm^2 carries a maximum tensile load of 20 kN. The modulus of elasticity for the material of the rod is 200 GPa. Determine the percentage strain when the rod is carrying its maximum load.   [0.1%]

19 A metal tube 1.75 m long carries a tensile load and the maximum stress in the tube must not exceed 50 MPa. Determine the extension of the tube when loaded if the modulus of elasticity for the material is 70 GPa.   [1.25 mm]

20 A piece of aluminium wire is 5 m long and has a cross-sectional area of 100 mm^2. It is subjected to increasing loads, the extension being recorded for each load applied. The results are:

Load (kN)	0	1.12	2.94	4.76	7.00	9.10
Extension (mm)	0	0.8	2.1	3.4	5.0	6.5

Draw the load/extension graph and hence determine the modulus of elasticity for the material of the wire. [70 GPa]

21 In an experiment to determine the modulus of elasticity of a sample of copper, a wire is loaded and the corresponding extension noted. The results are:

Load (kN)	0	20	34	72	94	120
Extension (mm)	0	0.7	1.2	2.5	3.3	4.2

Draw the load/extension graph and determine the modulus of elasticity of the sample if the mean diameter of the wire is 1.151 mm and its length is 4.0 m.
[110 GPa]

22 A tensile test is carried out on a specimen of mild steel of gauge length 40 mm and diameter 7.35 mm. The results are:

Load (kN)	0	10	17	25	30	34	37.5	38.5	36
Extension (mm)	0	0.05	0.08	0.11	0.14	0.20	0.40	0.60	0.90

At fracture the final length of the specimen is 40.90 mm. Plot the load/extension graph and determine (a) the modulus of elasticity for mild steel; (b) the stress at the limit of proportionality; (c) the ultimate tensile strength; (d) the percentage elongation. [(a) 202 GPa; (b) 707 MPa; (c) 907 MPa; (d) 2.25%]

23 In a tensile test on a zinc specimen of gauge length 100 mm and diameter 15 mm a load of 100 kN produced an extension of 0.666 mm. Determine (a) the stress induced (b) the strain; (c) Young's modulus of elasticity.
[(a) 566 MPa; (b) 0.006 66; (c) 85 GPa]

24 The results of a tensile test are:
Diameter of specimen 20 mm; gauge length 50 mm; load at limit of proportionality 80 kN; Extension at limit of proportionality 0.075 mm; maximum load 100 kN; final length at point of fracture 60 mm.
    Determine (a) Young's modulus of elasticity; (b) the ultimate tensile strength; (c) the stress at the limit of proportionality; (d) the percentage elongation.
[(a) 169.8 GPa; (b) 318.3 MPa; (c) 254.6 MPa; (d) 20%]

25 What is a tensile test? Make a sketch of a typical load/extension graph for a mild steel specimen to the point of fracture and mark on the sketch the following:
(a) the limit of proportionality; (b) the elastic limit; (c) the yield point.

26 An aluminium alloy specimen of gauge length 75 mm and of diameter 11.28 mm was subjected to a tensile test, with these results:

Load (kN)	0	2.0	6.5	11.5	13.6	16.0	18.0	19.0	20.5	19.0
Extension (mm)	0	0.012	0.039	0.069	0.080	0.107	0.133	0.158	0.225	0.310

The specimen fractured at a load of 19.0 kN. Determine (a) the modulus of elasticity of the alloy; (b) the percentage elongation. [(a) 125 GPa; (b) 0.413%]

7 Forces acting at a point

A. MAIN POINTS CONCERNED WITH FORCES ACTING AT A POINT

1   When forces are all acting in the same plane, they are called **coplanar**. When forces act at the same time and at the same point, they are called **concurrent forces**.

2   Force is a **vector quantity** and thus has both a magnitude and a direction. A vector can be represented graphically by a line drawn to scale in the direction of the line of action of the force. Vector quantities may be shown by using bold, lower case letters, thus **ab** in *Fig 1* represents a force of 5 newtons acting in a direction due east.

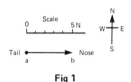

Fig 1

The resultant of two coplanar forces

3   For two forces acting at a point, there are three possibilities.

(a) For forces acting in the same direction and having the same line of action, the single force having the same effect as both of the forces, called the **resultant force** or just the **resultant**, is the arithmetic sum of the separate forces. Forces of F_1 and F_2 acting at point P, as shown in *Fig 2(a)* have exactly the same effect on point P as force F shown in *Fig 2(b)*, where $F = F_1 + F_2$ and acts in the same direction as F_1 and F_2. Thus, F is the resultant of F_1 and F_2.

(b) For forces acting in opposite directions along the same line of action, the resultant force is the arithmetic difference between the two

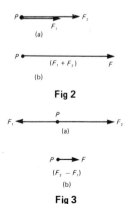

Fig 2

Fig 3

forces. Forces of F_1 and F_2 acting at point P as shown in *Fig 3(a)*, have exactly the same effect on point P as force F shown in *Fig 3(b)*, where $F = F_2 - F_1$ and acts in the direction of F_2, since F_2 is greater than F_1. Thus F is the resultant of F_1 and F_2 (see *Problem 1*).

(c) When two forces do not have the same line of action, the magnitude and direction of the resultant force may be found by a procedure called vector addition of forces. There are two graphical methods of performing **vector addition**, known as the triangle of forces method and the parallelogram of forces method.

The triangle of forces method:

(i) Draw a vector representing one of the forces, using an appropriate scale and in the direction of its line of action.

(ii) From the **nose** of this vector and using the same scale, draw a vector representing the second force in the direction of its line of action.

(iii) The resultant vector is represented in both magnitude and direction by the vector drawn from the tail of the first vector to the nose of the second vector. (See *Problems 2 and 3*).

The parallelogram of forces method:

(iv) Draw a vector representing one of the forces, using an appropriate scale and in the direction of its line of action.

(v) From the **tail** of this vector and using the same scale draw a vector representing the second force in the direction of its line of action.

(vi) Complete the parallelogram using the two vectors drawn in (iv) and (v) as two side of the parallelogram.

(vii) The resultant force is represented in both magnitude and direction by the vector corresponding to the diagonal of the parallelogram drawn from the tail of the vectors in (iv) and (v) (see *Problem 4*).

4   An alternative to the graphical methods of determining the resultant of two coplanar forces is by **calculation**. This can be achieved by trigonometry using the cosine rule and the sine rule (see *Problem 5*), or by resolution of forces (see para. 7).

The resultant of more than two coplanar forces

5   For the three coplanar forces F_1, F_2 and F_3 acting at a point as shown in *Fig 4*, the vector diagram is drawn using the nose-to-tail method of para. 3(c). The procedure is:

(i) Draw **oa** to scale to represent force F_1 in both magnitude and direction. (See *Fig 5*).

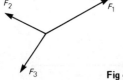

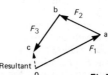

Fig 4   **Fig 5**

(ii) From the nose of **oa**, draw **ab** to represent force F_2.

(iii) From the nose of **ab**, draw **bc** to represent force F_3.

(iv) The resultant vector is given by length **oc** in *Figure 5*. The direction of resultant **oc** is from where we started, i.e. point o, to where we finished, i.e., point c. When acting by itself, the resultant force, given by **oc**, has the same effect on the point as forces F_1, F_2 and F_3 have when acting together. The resulting vector diagram of *Fig 5* is called the **polygon of forces** (see *Problems 6 and 7*).

6   When three or more coplanar forces are acting at a point and the vector diagram closes, there is no resultant. The forces acting at the point are in **equilibrium** (see *Problems 8 and 9*).

Resolution of forces

7   A vector quantity may be expressed in terms of its **horizontal** and **vertical** components. For example, a vector representing a force of 10 N at an angle of 60° to the horizontal is shown in *Fig 6*. If the horizontal line **oa** and the vertical line **ab** are

Fig 6

constructed as shown, then **oa** is called the horizontal component of the 10 N force and **ab** the vertical component of the 10 N force.

By trigonometry, $\cos 60° = \dfrac{oa}{ob}$. Hence the horizontal component, oa = 10 cos 60°.

Also, $\sin 60° = \dfrac{ab}{ob}$. Hence the vertical component, ab = 10 sin 60°.

This process is called **finding the horizontal and vertical components of a vector** or **the resolution of a vector**, and can be used as an alternative to graphical methods for calculating the resultant of two or more coplanar forces acting at a point.

For example, to calculate the resultant of a 10 N force acting at 60° to the horizontal and a 20 N force acting at −30° to the horizontal (see *Fig 7*) the procedure is as follows:

(i) Determine the horizontal and vertical components of the 10 N force, i.e. horizontal component, oa = 10 cos 60° = 5.0 N, and vertical component, ab = 10 sin 60° = 8.66 N.

(ii) Determine the horizontal and vertical components of the 20 N force, i.e. horizontal component,

od = 20 cos (−30°) = 17.32 N, and vertical component,

oc = 20 sin (−30°) = −10.0 N

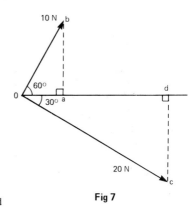

Fig 7

(iii) Determine the total horizontal component, i.e. oa + od = 5.0+17.32=22.32 N

(iv) Determine the total vertical component, i.e. ab + cd = 8.66+(−10.0)=−1.34 N.

(v) Sketch the total horizontal and vertical components as shown in *Fig 8*. The resultant of the two components is given by length **or** and, by Pythagoras' theorem,

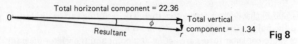

Total horizontal component = 22.36

Total vertical component = − 1.34

Resultant ϕ

Fig 8

or $= \sqrt{[(22.32)^2 + (1.34)^2]} = 22.36$ N, and using trigonometry,

angle $\phi = \arctan \dfrac{1.34}{22.32} = 3° \ 26'$.

Hence the resultant of the 10 N and 20 N forces shown in *Fig 7* is 22.36 N at an angle of $-3° \ 26'$ to the horizontal. This example demonstrates the use of resolution of forces for calculating the resultant of two coplanar forces acting at a point. However the method may be used for more than two forces acting at a point. (See *Problems 10 to 12*).

Summary

8 (a) To determine the resultant of two coplanar forces acting at a point, four methods are commonly used. They are:

by drawing: (1) triangle of forces method; (2) parallelogram of forces method, and

by calculation: (3) use of cosine and sine rules; (4) resolution of forces.

(b) To determine the resultant of more than two coplanar forces acting at a point, two methods are commonly used. They are:

by drawing: (1) polygon of forces method; and

by calculation: (2) resolution of forces.

B. WORKED PROBLEMS ON FORCES ACTING AT A POINT

Problem 1 Determine the resultant of forces of 5 kN and 8 kN, (a) acting in the same direction and having the same line of action; (b) acting in opposite directions but having the same line of action.

(a) The vector diagram of the two forces acting in the same direction is shown in *Fig 9(a)*, which assumes that the line of action is horizontal although, since it is not specified could be in any direction. From para. 3(a), the resultant force F is given by:

$F = F_1 + F_2$

i.e. $F = (5+8)$kN $= \mathbf{13 \ kN}$ in the direction of the original forces.

Fig 9

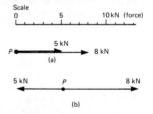

(b) The vector diagram for the two forces acting in opposite directions is shown in *Fig 9(b)*, again assuming that the line of action is in a horizontal direction. From para. 3(b), the resultant force F is given by:

$F = F_2 - F_1$

i.e. $F = (8 - 5)$kN $= \mathbf{3 \ kN}$ in the direction of the 8 kN force.

Problem 2 Determine the magnitude and direction of the resultant of a force of 15N acting horizontally to the right and a force of 20N, inclined at an angle of 60° to the 15 N force. Use the triangle of forces method.

Using the procedure given in para. 3(c) and with reference to *Fig 10*

(i) **ab** is drawn 15 units long horizontally;

(ii) from b, **bc** is drawn 20 units long, inclined at an angle of 60° to ab. (Note, in angular measure, an angle of 60° from ab means 60° in an anticlockwise direction.)

(iii) By measurement, the resultant **ac** is 30.5 units long inclined at an angle of 35° to ab.

That is, the resultant force is **30.5 N** inclined at an angle of **35°** to the 15 N force.

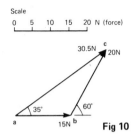

Fig 10

Problem 3 Find the magnitude and direction of the two forces given, using the triangle of forces method.

First force: 1.5 kN acting at an angle of 30°.
Second force: 3.7 kN acting at an angle of −45°.

From the procedure given in para. 3(c) and with reference to *Fig 11*:

(i) **ab** is drawn at an angle of 30° and 1.5 units in length.

(ii) From b, **bc** is drawn at an angle of −45° and 3.7 units in length. (Note, an angle of −45° means a clockwise rotation of 45° from a line drawn horizontally to the right.)

(iii) By measurement, the resultant **ac** is 4.3 units long at an angle of −25°. That is, the resultant force is **4.3 kN** at an angle of −25°.

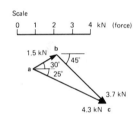

Fig 11

Problem 4 Use the parallelogram of forces method to find the magnitude and direction of the resultant of a force of 250 N acting at an angle of 135° and a force of 400 N acting at an angle of −120°.

From the procedure given in para. 3(c) and with reference to *Fig 12*:

(iv) **ab** is drawn at an angle of 135° and 250 units in length;

(v) **ac** is drawn at an angle of −120° and 400 units in length;

(vi) bc and cd are drawn to complete the parallelogram;

(vii) **ad** is drawn. By measurement **ad** is 413 units long at an angle of −156°. That is, the resultant force is **413 N** at an angle of **−156°**.

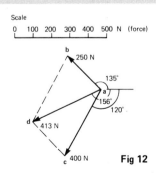

Fig 12

91

Problem 5 Use the cosine and sine rules to determine the magnitude and direction of the resultant of a force of 8 kN acting at an angle of 50° to the horizontal and a force of 5 kN acting at an angle of −30° to the horizontal.

The space diagram is shown in *Fig 13(a)*. A sketch is made of the vector diagram, **oa** representing the 8 kN force in magnitude and direction and **ab** represents the

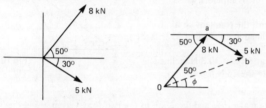

(a) space diagram   (b) vector diagram   **Fig 13**

5 kN force in magnitude and direction. The resultant is given by length **ob**. By the cosine rule, ob² = oa² + ab² − 2(oa)(ab) cos ∠oab
$$= 8^2 + 5^2 − 2(8)(5) \cos 100°, \text{ since } ∠oab = 180° − 50° − 30° = 100°$$
$$= 64 + 25 − (−13.892) = 102.892$$
Hence ob = √(102.892) = 10.14 kN

By the sine rule, $\dfrac{5}{\sin ∠aob} = \dfrac{10.14}{\sin 100°}$

from which, sin ∠aob = $\dfrac{5 \sin 100°}{10.14}$ = 0.4856

Hence ∠aob = arcsin (0.4856) = 29° 3'.
Thus angle ϕ in *Fig 13(b)* is 50° − 29° 3' = 20° 57':
Hence the resultant of the two forces is 10.14 kN acting at an angle of 20° 57' to the horizontal.

Problem 6 Determine graphically the magnitude and direction of the resultant of these three coplanar forces, which may be considered as acting at a point.
Force A, 12 N acting horizontally to the right; force B, 7N inclined at 60° to force A; force C, 15 N inclined at 150° to force A.

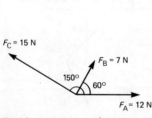

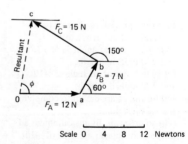

Fig 14   **Fig 15**

The space diagram is shown in *Fig 14*.

The vector diagram (*Fig 15*) is produced as follows:

(i) **oa** represents the 12 N force in magnitude and direction.

(ii) From the nose of **oa**, **ab** is drawn inclined at 60° to oa and 7 units long.

(iii) From the nose of **ab**, **bc** is drawn 15 units long inclined at 150° to oa (i.e. 150° to the horizontal).

(iv) **oc** represents the resultant. By measurement, the resultant is 13.8 N inclined at 80° to the horizontal.

Thus the resultant of the three forces, F_A, F_B and F_C is a force of 13.8 N at 80° to the horizontal.

Problem 7 The following coplanar forces are acting at a point, the given angles being measured from the horizontal: 100 N at 30°, 200 N at 80°, 40 N at −150°, 120 N at −100° and 70 N at −60° Determine graphically the magnitude and direction of the resultant of the five forces.

The five forces are shown in the space diagram of *Fig 16*. Since the 200 N and 120 N forces have the same line of action but are in opposite sense, they can be represented by a single force of 200−120, i.e. 80 N acting at 80° to the horizontal. Similarly, the 100 N and 40 N forces can be represented by a force of 100−40, i.e. 60 N acting at 30° to the horizontal. Hence the space diagram of *Fig 16* may be represented by the space diagram of *Fig 17*. Such a simplification of the vectors is not essential but it is easier to construct the vector diagram from a space diagram having three forces, than one with five. The vector diagram shown in *Fig 18*, **oa** representing the 60 N force, **ab** representing the 80 N force and **bc** the 70 N force. The resultant, **oc**, is found by measurement to represent a force of 112 N and angle ϕ is 25°.

Thus the five forces shown in *Fig 16* may be represented by a single force of 112 N at 25° to the horizontal.

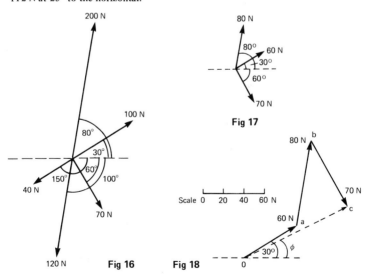

Fig 17

Fig 16

Fig 18

Scale 0   20   40   60 N

93

Problem 8 A load of 200 N is lifted by two ropes connected to the same point on the load, making angles of 40° and 35° with the vertical. Determine graphically the tensions in each rope when the system is in equilibrium.

The space diagram is shown in *Fig 19*. Since the system is in equilibrium, the vector diagram must close. The vector diagram (*Fig 20*) is drawn as follows:

(i) The load of 200 N is drawn vertically as shown by **oa**.

(ii) The direction only of force F_1 is known, so from point a, ad is drawn at 40° to the vertical.

(iii) The direction only of force F_2 is known, so from point o, oc is drawn at 35° to the vertical.

(iv) Lines ad and oc cross at point b. Hence the vector diagram is given by triangles oab. By measurement, ab is 119 N and **ob** is 133 N.

Thus the tensions in the ropes are
$F_1 = 119$ N and $F_2 = 133$ N

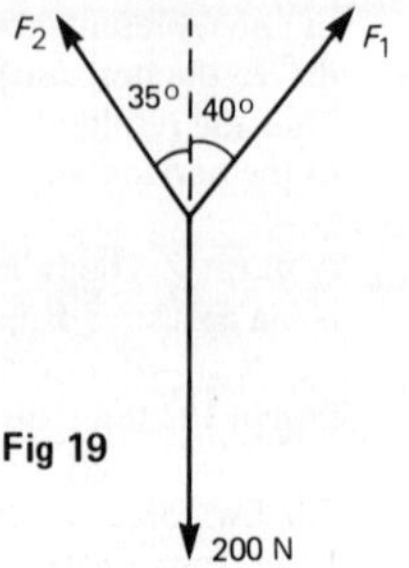

Fig 19

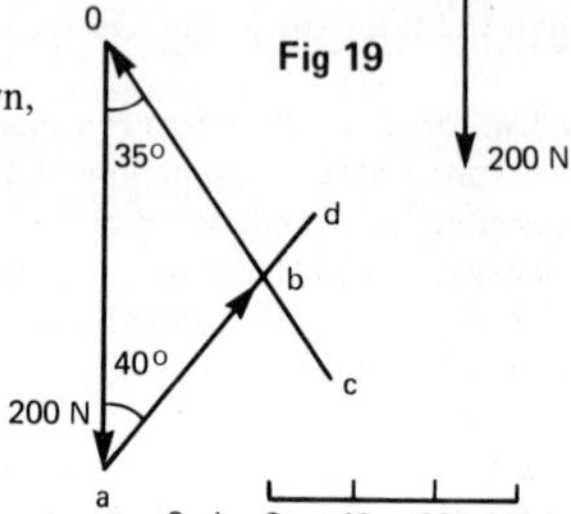

Fig 20   Scale 0   40   80   120 N

Problem 9 Five coplanar forces are acting on a body and the body is in equilibrium. The forces are: 12 kN acting horizontally to the right, 18 kN acting at an angle of 75°, 7 kN acting at an angle of 165°, 16 kN acting from the nose of the 7 kN force, and 15 kN acting from the nose of the 16 kN force. Determine the directions of the 16 kN and 15 kN forces relative to the 12 kN force.

With reference to *Fig 21*, oa is drawn 12 units long horizontally to the right. From point a, *ab* is drawn 18 units long at an angle of 75°. From b, *bc* is drawn 7 units long at an angle of 165°. The direction of the 16 kN force is not known, thus arc pq is drawn iwth a compass, with centre at c, radius 16 units. Since the forces are at equilibrium, the polygon of forces must close. Using a compass with centre at 0, arc rs is drawn having a radius 15 units. The point where the arcs intersect is at d. By measurement, angle $\phi = 198°$ and $\alpha = 291°$.

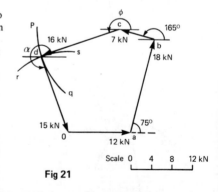

Fig 21   Scale 0   4   8   12 kN

Thus the 16 kN force acts at an angle of 198° (or −162°) to the 12 kN force and the 15 kN force acts at an angle of 291° (or −69°) to the 12 kN force.

Problem 10 Forces of 5.0 N at 25° and 8.0 N at 112° act at a point. By resolving these forces into horizontal and vertical components, determine their resultant.

The space diagram is shown in *Fig 22*.
(i) The horizontal component of the
5.0 N force, oa = 5.0 cos 25° = 4.532;
the vertical component of the 5.0 N
force, ab = 5.0 sin 25° = 2.113.
(ii) The horizontal component of the
8.0 N force, oc = 8.0 cos ∠cod
= 8.0 cos 68° = 2.997. However, in
the second quadrant, the cosine of an
acute angle is negative, hence the
horizontal component of the 8.0 N
force is −2.997. . The vertical component of the 8.0 N force,
cd = 8.0 sin ∠cod = 8.0 sin 68° = 7.417.
Since, in the second quadrant, the
sine of an acute angle is positive, the
vertical component of the 8.0 N force
is +7.417.

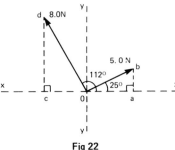

Fig 22

(A useful check is that the vertical components ab and cd are both above the XX axis and are thus both positive; horizontal component oa is positive since it is to the right of axis YY and the horizontal component oc is negative since it is to the left of axis YY).

With a calculator, horizontal and vertical components can be determined easily. For example, component oc in *Fig 22* is equivalent o 8.0 cos 112°, which is evaluated as −2.997 directly with a calculator without having to consider the equivalent acute angle (i.e. ∠cod), and whether it is positive or negative in that quadrant.
(iii) Total horizontal component = oa + oc = 4.532 + (−2.997) = +1.535
(iv) Total vertical component = ab + dc = 2.113 + 7.417 = +9.530
(v) The components are shown sketched in
Fig 23. By Pythagoras' theorem,
$r = \sqrt{[(1.535)^2 + (9.530)^2]} = 9.653$,

and by trigonometry, angle ϕ = arctan $\dfrac{9.530}{1.535}$
= 80° 51′.

Hence the resultant of the two forces shown in
Figure 22 is a force of 9.653 N acting at 80° 51′
to the horizontal.

Fig 23

Problem 11 Determine by resolution of forces the resultant of the following three coplanar forces acting at a point: 200 N acting at 20° to the horizontal; 400 N acting at 165° to the horizontal; 500 N acting at 250° to the horizontal.

A tabular approach using a calculator may be made as shown below.

	Horizontal component	Vertical component
Force 1	200 cos 20° = 187.94	200 sin 20° = 68.40
Force 2	400 cos 165° = −386.37	400 sin 165° = 103.53
Force 3	500 cos 250° = −171.01	500 sin 250° = −469.85
	Total horizontal component = −369.44	Total vertical component = −297.92

The total horizontal and vertical components are shown in *Fig 24*.

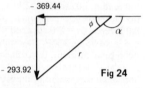

Resultant $r = \sqrt{[(369.44)^2 + (297.92)^2]}$
$= 474.60$

and angle $\phi = \arctan \dfrac{297.92}{369.44} = 38° \, 53'$,

from which, $\alpha = 180° - 38°53' = 141° \, 7'$.

Fig 24

Thus the resultant of the three forces given is 474.6 N acting at an angle of −141° 7′ (or +218° 53′) to the horizontal.

Problem 12 The following coplanar forces act at a point: Force A is 18 kN at 15° to the horizontal; force B is 25 kN at 126° to the horizontal; force C is 10 kN at 197° to the horizontal; force D is 15 kN at 246° to the horizontal; force E is 30 kN at 331° to the horizontal.
Determine the resultant of the five forces by resolution of forces.

Using a tabular approach:

	Horizontal component	Vertical component
Force A	18 cos 15° = 17.39	18 sin 15° = 4.66
Force B	25 cos 126° = −14.69	25 sin 126° = 20.23
Force C	10 cos 197° = −9.56	10 sin 197° = −2.92
Force D	15 cos 246° = −6.10	15 sin 246° = −13.70
Force E	30 cos 331° = 26.24	30 sin 331° = −14.54
	Total horizontal component = +13.28	Total vertical component = −6.27

The total horizontal and vertical components are shown in *Fig 25*.
By Pythagoras' theorem, resultant
$r = \sqrt{[(13.28)^2 + (6.27)^2]} = 14.69$

and angle $\phi = \arctan \dfrac{6.27}{13.28} = 25° \, 16'$.

Hence the resultant of the five forces is 14.69 kN acting at an angle of −25° 16′.

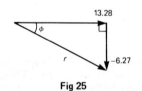

Fig 25

C. FURTHER PROBLEMS ON FORCES ACTING AT A POINT

(a) SHORT ANSWER PROBLEMS

1 State the meaning of the term 'coplanar'.

2 What is a concurrent force?

3   State what is meant by a triangle of forces.

4   State what is meant by a parallelogram of forces.

5   State what is meant by a polygon of forces.

6   When a vector diagram is drawn representing coplanar forces acting at a point, and there is no resultant, the forces are in . . . .

7   Two forces of 6 N and 9 N act horizontally to the right. The resultant is . . . . . . N acting . . . .

8   A force of 10 N acts at an angle of 50° and another force of 20 N acts at an angle of 230°. The resultant is a force . . . . N acting at an angle of . . . . °.

9   What is meant by 'resolution of forces'?

10 A coplanar force system comprises a 20 kN force acting horizontally to the right, 30 kN at 45°, 20 kN at 180° and 25 kN at 225°. The resultant is a force of . . . . . . N acting at an angle of . . . . ° to the horizontal.

(b) MULTI-CHOICE PROBLEMS (answers on page 172)

1   The magnitude of the resultant of the vectors shown in *Fig 26* is:
    (a) 2 N; (b) 12 N; (c) 35 N; (d) −2 N.

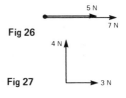

Fig 26

2   The magnitude of the resultant of the vectors shown in *Fig 27* is:
    (a) 7 N; (b) 5 N; (c) 1 N; (d) 12 N.

Fig 27

3   Which of the following statements is false?
    (a) There is always a resultant vector required to close a vector diagram representing a system of coplanar forces acting at a point, which are not in equilibrium.
    (b) A vector quantity has both magnitude and direction.
    (c) A vector diagram representing a system of coplanar forces acting at a point when in equilibrium does not close.
    (d) Concurrent forces are those which act at the same time at the same point.

4   Which of the following statements is false?
    (a) The resultant of coplanar forces of 1 N, 2 N and 3 N acting at a point can be 4 N.
    (b) The resultant of forces of 6 N and 3 N acting in the same line of action but opposite in sense is 3 N.
    (c) The resultant of forces of 6 N and 3 N acting in the same sense and having the same line of action is 9 N.
    (d) The resultant of coplanar forces of 4 N at 0°, 3 N at 90° and 8 N at 180° is 15 N.

5   A space diagram for a force system is shown in *Fig 28*. Which of the vector diagrams in *Fig 29* does *not* represent this force system?

6   With reference to *Fig 30*, which of the following statements is false?
    (a) The horizontal component of F_A is 8.66 N.
    (b) The vertical component of F_B is 10 N.
    (c) The horizontal component of F_C is 0.
    (d) The vertical component of F_D is 4 N.

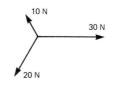

Fig 28

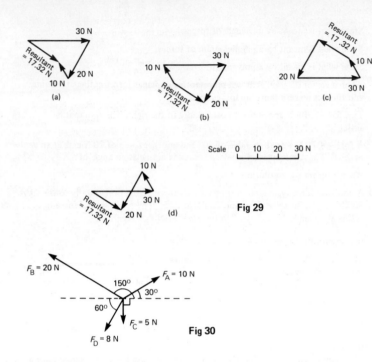

Fig 29

Fig 30

(c) CONVENTIONAL PROBLEMS

In *Problems 1 to 10*, use a graphical method to determine the magnitude and direction of the resultant of the forces given.

1   1.3 kN and 2.7 kN, having the same line of action and acting in the same direction.
[4.0 kN in the direction of the forces]

2   470 N and 538 N having the same line of action but acting in opposite directions.
[68 N in the direction of the 538 N force]

3   13 N at 0° and 25 N at 30°.   [36.8 N at 20°]

4   5 N at 60° and 8 N at 90°.   [12.6 N at 79°]

5   1.3 kN at 45° and 2.8 kN at −30°.   [3.4 kN at −8°]

6   1.7 N at 45° and 2.4 N at −60°.   [2.6 kN at −20°]

7   9 N at 126° and 14 N at 223°.   [15.7 N at −172°]

8   23.8 N at −50° and 14.4 N at 215°.   [26.7 N at −82°]

9   0.7 kN at 147° and 1.3 kN at −71°.   [0.86 kN at −100°]

10 47 N at 79° and 58 N at 247°.   [15.5 N at −152°]

11 Resolve a force of 23.0 N at an angle of 64° into its horizontal and vertical components.   [10.08 N; 20.67 N]

98

12 Forces of 7.6 kN at 32° and 11.8 kN at 143° act at a point. Use the cosine and sine rules to calculate the magnitude and direction of their resultant.

[11.52 kN at 104° 59']

13 In *Problems 3 to 10*, calculate the resultant of the given forces by using the cosine and sine rules.

14 Forces of 5 N at 21° and 9 N at 126° act at a point. By resolving these forces into horizontal and vertical components, determine their resultant. [9.09 N at 93° 55']

In *Problems 15 to 17*, determine graphically the magnitude and direction of the resultant of the coplanar forces given which are acting at a point.

15 Force A, 12 N acting horizontally to the right; force B, 20 N acting at 140° to force A; force C, 16 N acting at 290° to force A. [3.06 N at −45° to force A]

16 Force 1, 23 kN acting at 80° to the horizontal; force 2, 30 kN acting at 37° to force 1; force 3, 15 kN acting at 70° to force 2.

[53.5 kN at 37° to force 1 (i.e. 117° to the horizontal)]

17 Force P, 50 kN acting horizontally to the right; force Q, 20 kN at 70° to force P; force R, 40 kN at 170° to force P; force S, 80 kN at 300° to force P.

[72 kN at −37° to force P]

18 Four horizontal wires are attached to a telephone pole and exert tensions of 30 N to the south, 20 N to the east, 50 N to the north-east and 40 N to the north-west. Determine the resultant force on the pole and its direction.

[43.18 N, 38° 49' east of north]

19 Four coplanar forces acting on a body are such that it is in equilibrium. The vector diagram for the forces is such that the 60 N force acts vertically upwards, the 40 N force acts at 65° to the 60 N force, the 100 N force acts from the nose of the 60 N force and the 90 N force acts from the nose of the 100 N force. Determine the direction of the 100 N and 90 N forces relative to the 60 N force.

$$\begin{bmatrix} 100 \text{ N force at } 263° \text{ to the } 60 \text{ N force;} \\ 90 \text{ N force at } 132° \text{ to the } 60 \text{ N force} \end{bmatrix}$$

20 A load of 12.5 N is lifted by two strings connected to the same point on the load, making angles of 22° and 31° on opposite sides of the vertical. Determine the tensions in the strings. [5.8 N; 8.0 N]

21 Determine, by resolution of forces, the resultant of the following three coplanar forces acting at a point: 10 kN acting at 32° to the horizontal; 15 kN acting at 170° to the horizontal; 20 kN acting at 240° to the horizontal.

[18.82 kN at 210° 2' to the horizontal]

22 In *Problems 15 to 17*, calculate the resultant force in each case by resolution of the forces.

23 The following coplanar forces act at a point: Force A, 15 N acting horizontally to the right; force B, 23 N at 81° to the horizontal; force C, 7 N at 210° to the horizontal; force D, 9 N at 265° to the horizontal; force E, 28 N at 324° to the horizontal. Determine the resultant of the five forces by resolution of the forces.

[34.96 N at −10° 14' to the horizontal]

24 Forces of 5 kN, 3 kN, 7 kN and *F* are coplanar and act at a point on a body. Their directions are 63°, 125°, 302° and ϕ° respectively from the horizontal. Use resolution

of forces to determine the values of F and ϕ when the forces are in equilibrium.

$$[F = 4.37 \text{ kN}; \phi = 192° 55']$$

25 A two-legged sling and hoist chain used for lifting machine parts is shown in *Fig 31*. Determine the forces in each leg of the sling if parts exerting a downward force of 15 kN are lifted.   [9.96 kN; 7.77 kN]

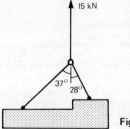

Fig 31

8 Simply supported beams

A. MAIN POINTS CONCERNED WITH SIMPLY SUPPORTED BEAMS

1   When using a spanner to tighten a nut, a force tends to turn the nut in a clockwise direction. This turning effect of a force is called the **moment of a force** or more briefly, just a **moment**. The size of the moment acting on the nut depends on two factors:
(a) the size of the force acting at right angles to the shank of the spanner, and
(b) the perpendicular distance between the point of application of the force and the centre of the nut.

In general, with reference to *Fig 1*,
the moment M of a force acting at a point
P is force $\times$ perpendicular distance between
the line of action of the force and P.
i.e., $M = F \times d$

Fig 1

The unit of a moment is the newton metre, (Nm). Thus, if force F in *Figure 1* is 7 N
and distance d is 3 m, then the moment P is $7(N) \times 3(m)$, i.e. 21 Nm.
(See *Problems 1 and 2*).

2   If more than one force is acting on an object and the forces do not act at a point,
then the turning effect of the forces, that is, the moment of the forces, must be
considered.

Fig 2 shows a beam with its support,
(known as its pivot or fulcrum), at P, acting
vertically upwards, and forces F_1 and F_2
acting vertically downwards at distances
a and b respectively from the fulcrum.
A beam is said to be in **equilibrium** when
there is no tendency for it to move.

Fig 2

There are two conditions for equilibrium:
(i) The sum of the forces acting vertically downwards must be equal to the sum of
the forces acting vertially upwards, i.e. for *Fig 2*, $R_P = F_1 + F_2$
(ii) The total moment of the forces acting on a beam must be zero; for the total
moment to be zero:
    'the sum of the clockwise moments about any point must be equal to the sum of
    the anticlockwise moments about that point'.

This statement is known as the **principle of moments**. Hence, taking moments about

101

P in *Fig 2*, $F_2 \times b$ = the clockwise moment, and $F_1 \times a$ = the anticlockwise moment. Thus for equilibrium:

$$F_1 a = F_2 b$$

(See *Problems 3 to 5*).

3 (i) A **simply supported beam** is one which rests on two supports and is free to move horizontally.

(ii) Two typical simply supported beams having loads acting at given points on the beam, (called **point loading**), are shown in *Fig 3*. A man whose mass exerts a force F vertically downwards, standing on a wooden plank which is simply supported at its ends, may, for example, be represented by the beam diagram of *Fig 3(a)* if the mass of the plank is neglected. The forces exerted by the supports on the plank, R_P and R_Q, act vertically upwards, and are called **reactions**.

(iii) When the forces acting are all in one plane, the algebraic sum of the moments can be taken about **any** point.

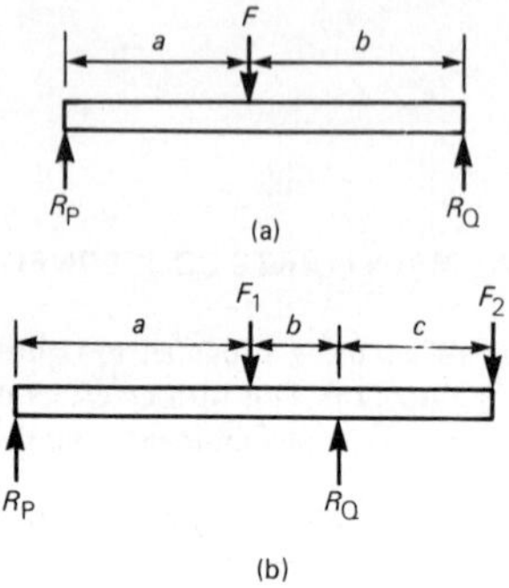

For the beam in *Fig 3(a)* at equilibrium:

(i) $R_P + R_Q = F$

and (ii) taking moments about R_P, $F_a = R_Q b$
(Alternatively, taking moments about F, $R_P a = R_Q b$)
For the beam in *Fig 3(b)*, at equilibrium:

(i) $R_P + R_Q = F_1 + F_2$

and (ii) taking moments about R_Q, $R_P(a + b) + F_2 c = F_1 b$.

(iv) Typical practical applications of simply supported beams with point loadings include bridges, beams in buildings, and beds of machine tools.

(See *Problems 6 to 10*).

Fig 3

B. WORKED PROBLEMS ON SIMPLY SUPPORTED BEAMS

Problem 1 A force of 15 N is applied to a spanner at an effective length of 140 mm from the centre of a nut. Calculate (a) the moment of the force applied to the nut; (b) the magnitude of the force required to produce the same moment if the effective length is reduced to 100 mm.

From para. 1, $M = F \times d$, where M is the turning moment, F is the force applied at right angles to the spanner and d is the effective length between the force and the centre of the nut. Thus, with reference to *Fig 4(a)*.

(a) Turning moment, $M = 15 \text{ N} \times 140 \text{ mm}$
    $= 2100 \text{ N mm}$

    $= 2100 \text{ N mm} \times \dfrac{1 \text{ m}}{1000 \text{ mm}}$

    $= 2.1 \text{ Nm}$

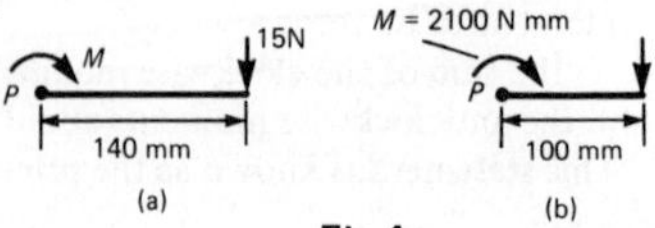

Fig 4

(b) Turning moment, M is 2100 N mm and the effective length d becomes 100 mm, (see *Fig 4(b)*). Applying $M = F \times d$ gives:

2100 N mm $= F \times 100$ mm

from which, force $F = \dfrac{2100 \text{ N mm}}{100 \text{ mm}} = \textbf{21 N}$

Problem 2 A moment of 25 Nm is required to operate a lifting jack. Determine the effective length of the handle of the jack if the force applied to it is (a) 125 N; (b) 0.4 kN.

From para. 1, moment $M = F \times d$, where F is the force applied at right-angles to the handle and d is the effective length of the handle. Thus:

(a) 25 Nm $= 125$ N $\times d$, from which,

effective length $d = \dfrac{25 \text{ Nm}}{125 \text{ N}} = \dfrac{1}{5} \text{m} = \dfrac{1000}{5}$ mm

$= \textbf{200 mm}$

(b) Turning moment M is 25 Nm and the force F becomes 0.4 kN, i.e., 400 N. Since $M = F \times d$, then 25 Nm $= 400$ N $\times d$

Thus, the effective length, $d = \dfrac{25 \text{ Nm}}{400 \text{ N}} = \dfrac{1}{16} \text{ m} = \textbf{62.5 mm}$

Problem 3 A system of forces is as shown in *Fig 5*.
(a) If the system is in equilibrium find the distance d.
(b) If the point of application of the 5 N force is moved to point **P**, distance 200 mm from the support, find the new value of F to replace the 5 N force for the system to be in equilibrium.

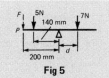

Fig 5

From para. 2,
(a) The clockwise moment M_1 is due to a force of 7 N acting at a distance d from the support, called the **fulcrum**, i.e.

$$M_1 = 7 \text{ N} \times d$$

The anticlockwise moment M_2 is due to a force of 5 N acting at a distance of 140 mm from the fulcrum, i.e.

$$M_2 = 5 \text{ N} \times 140 \text{ mm}.$$

Applying the principle of moments, for the system to be in equilibrium:

clockwise moment $=$ anticlockwise moment
i.e. 7 N $\times d = 5 \times 140$ N mm

Hence, distance $d = \dfrac{5 \times 140 \text{ N mm}}{7 \text{ N}} = \textbf{100 mm}$

(b) When the 5 N force is replaced by force F at a distance of 200 mm from the fulcrum, the new value of the anticlockwise moment is $F \times 200$. For the system to be in equilibrium:

clockwise moment $=$ anticlockwise moment
i.e. (7×100)N mm $= F \times 200$ mm

Hence, new value of force, $F = \dfrac{700 \text{ N mm}}{200 \text{ mm}} = \textbf{3.5 N}$

Problem 4 A beam is supported at its centre on a fulcrum and forces act as shown in *Fig 6*. Calculate (a) force F for the beam to be in equilibrium; (b) the new position of the 23 N force when F is decreased to 21 N, if equilibrium is to be maintained.

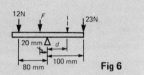

Fig 6

From para. 2:

(a) The clockwise moment, M_1 is due to the 23 N force acting at a distance of 100 mm from the fulcrum, i.e.

$M_1 = 23 \times 100 = 2300$ N mm

There are two forces giving the anticlockwise moment M_2. One is the force F acting at a distance of 20 mm from the fulcrum and the other a force 12 N acting at a distance of 80 mm.

Thus $M_2 = (F \times 20) + (12 \times 80)$ N mm

Applying the principle of moments:

clockwise moment = anticlockwise moments

i.e. $2300 = (F \times 20) + (12 \times 80)$

Hence $F \times 20 = 2300 - 960$

i.e. force $F = \dfrac{1340}{20} = \mathbf{67}$ **N**

(b) The clockwise moment is now due to a force of 23 N acting at a distance of, say, d, from the fulcrum. Since the value of F is decreased to 21 N, the anticlockwise moment is $(21 \times 20) + (12 \times 80)$ N mm.

Applying the principle of moments, $23 \times d = (21 \times 20) + (12 \times 80)$

i.e. distance $d = \dfrac{420 + 960}{23} = \dfrac{1380}{23} = \mathbf{60}$ **mm**

Problem 5 For the centrally supported uniform beam shown in *Fig 7* determine the values of forces F_1 and F_2 when the beam is in equilibrium.

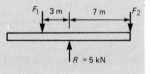

Fig 7

At equilibrium: (i) $R = F_1 + F_2$, i.e. $5 = F_1 + F_2$ (1)

and   (ii) $F_1 \times 3 = F_2 \times 7$ (2)

From equation (1), $F_2 = 5 - F_1$

Substituting for F_2 in equation (2) gives: $F_1 \times 3 = (5 - F_1) \times 7$, i.e. $3F_1 = 35 - 7F_1$
$10F_1 = 35$, from which $F_1 = 3.5$ kN

Since $F_2 = 5 - F_1$, $F_2 = 1.5$ kN.

Thus at equilibrium, force $F_1 = 3.5$ kN and force $F_2 = 1.5$ kN

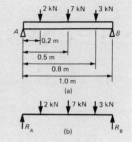

Problem 6 A beam is loaded as shown in *Fig 8*. Determine (a) the force acting on the beam support at B; (b) the force acting on the beam support at A, neglecting the mass of the beam.

Fig 8

A beam supported as shown in *Fig 8* is called a **simply supported beam**.

(a) Taking moments about point A and applying the principle of moments gives:

clockwise moments = anticlockwise moments

$(2 \times 0.2) + (7 \times 0.5) + (3 \times 0.8)$ kN m $= R_B \times 1.0$ m, where R_B is the force supporting the beam at B, as shown in *Figure 8(b)*
Thus $(0.4 + 3.5 + 2.4)$ kN m $= R_B \times 1.0$ m

i.e. $R_B = \dfrac{6.3 \text{ kN m}}{1.0 \text{ m}} = \textbf{6.3 kN}$

(b) For the beam to be in equilibrium, the forces acting upwards must be equal to the forces acting downwards, thus $R_A + R_B = (2 + 7 + 3)$ kN
$R_B = 6.3$ kN, thus $R_A = 12 - 6.3 = \textbf{5.7 kN}$

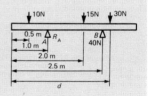

Problem 7 For the beam shown in *Fig 9* calculate (a) the force acting on support A; (b) distance d, neglecting any forces arising from the mass of the beam.

Fig 9

(a) From para. 2, (the forces acting in an upward direction) = (the forces acting in a downward direction)

Hence, $(R_A + 40)$ N $= (10 + 15 + 30)$ N

$R_A = 10 + 15 + 30 - 40 = \textbf{15 N}$

(b) Taking moments about the left-hand end of the beam and applying the principle of moments gives:

clockwise moments = anticlockwise moments

$(10 \times 0.5) + (15 \times 2.0)$ N m $+ 30$ N $\times d = (15 \times 1.0) + (40 \times 2.5)$ N m
i.e. 35 Nm $+ 30$ N $\times d = 115$ Nm,

from which, distance $d = \dfrac{(115 - 35) \text{ Nm}}{30 \text{ N}} = 2\frac{2}{3}$ **m**

105

Problem 8 A metal bar AB is 4.0 m long and is supported at each end in a horizontal position. It carries loads of 2.5 kN and 5.5 kN at distances of 2.0 m and 3.0 m respectively from A. Neglecting the mass of the beam, determine the reactions of the supports when the beam is in equilibrium.

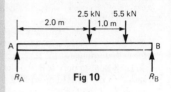

Fig 10

The beam and its loads are shown in *Fig 10*

At equilibrium, $R_A + R_B = 2.5 + 5.5 = 8.0$ kN   (1)

Taking moments about A, clockwise moments = anticlockwise moment

i.e. $(2.5 \times 2.0) + (5.5 \times 3.0) = 4.0 R_B$, or $5.0 + 16.5 = 4.0 R_B$

from which, $R_B = \dfrac{21.5}{4.0} = 5.375$ kN

From equation (1), $R_A = 8.0 - 5.375 = 2.625$ kN.

Thus the reactions at the supports at equilibrium are 2.625 kN at A and 5.375 kN at B.

Problem 9 A beam PQ is 5.0 m long and is supported at its ends in a horizontal position as shown in *Fig 11*. Its mass is equivalent to a force of 400 N acting at its centre as shown. Point loads of 12 kN and 20 kN act on the beam in the positions shown. When the beam is in equilibrium, determine (a) the reactions of the supports, R_P and R_Q, and (b) the position to which the 12 kN load must be moved for the force on the supports to be equal.

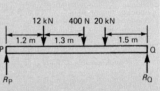

Fig 11

(a) At equilibrium, $R_P + R_Q = 12 + 0.4 + 20 = 32.4$ kN   (1)

Taking moments about P: clockwise moments = anticlockwise moment

i.e. $(12 \times 1.2) + (0.4 \times 2.5) + (20 \times 3.5) = (R_Q \times 5.0)$

$14.4 + 1.0 + 70.0 = 5.0 R_Q$

from which, $R_Q = \dfrac{85.4}{5.0} = 17.08$ kN

From equation (1), $R_P = 32.4 - R_Q = 32.4 - 17.08 = 15.32$ kN.

(b) For the reactions of the supports to be equal, $R_P = R_Q = \dfrac{32.4}{2} = 16.2$ kN

Let the 12 kN load be at a distance d metres from P (instead of at 1.2 m from P). Taking moments about point P gives:

$12 d + (0.4 \times 2.5) + (20 \times 3.5) = 5.0 R_Q$

i.e. $12d + 1.0 + 70.0 = 5.0 \times 16.2$, and $12d = 81.0 - 71.0$

from which, $d = \dfrac{10.0}{12} = 0.833$ m

Hence the 12 kN load needs to be moved to a position 833 mm from P for the reactions of the supports to be equal (i.e. 367 mm to the left of its original position)

106

Problem 10 A uniform steel girder AB which is 6.0 m long has a mass equivalent to 4.0 kN acting at its centre. The girder rests on two supports at C and B as shown in *Fig 12*. A point load of 20.0 kN is attached to the beam as shown. Determine the value of force *F* which causes the beam to just lift off the support B.

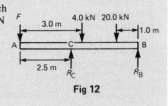

Fig 12

At equilibrium, $R_C + R_B = F + 4.0 + 20.0$
When the beam is just lifting off of the support B, then $R_B = 0$.
Hence $R_C = (F + 24.0)$ kN.
Taking moments about A: clockwise moments = anticlockwise moments
i.e. $(4.0 \times 3.0) + (5.0 \times 20.0) = (R_C \times 2.5) + (R_B \times 6.0)$
$12.0 + 100.0 = (F + 24.0) \times 2.5 + 0$
$\dfrac{112.0}{2.5} = (F + 24.0)$
from which, $F = 44.8 - 24.0 = 20.8$ kN,
i.e. **the value of force F which causes the beam to just lift off the support B is 20.8 kN.**

C. FURTHER PROBLEMS ON SIMPLY SUPPORTED BEAMS

(a) SHORT ANSWER PROBLEMS

1 The moment of a force is the product of . . . . and . . . .

2 When a beam has no tendency to move it is in . . . .

3 State the two conditions for equilibrium of a beam.

4 What is meant by a simply supported beam?

5 State two practical applications of simply supported beams.

(b) MULTI-CHOICE PROBLEMS (answers on page 172)

1 A force of 10 N is applied at right-angles to the handle of a spanner, 5 m from the centre of a nut. The moment on the nut is:
(a) 50 Nm; (b) 2 N/m; (c) 0.5 m/N; (d) 15 Nm.

2 The distance *d* in *Fig 13* when the beam is in equilibrium is:
(a) 0.5 m; (b) 1.0 m; (c) 4.0 m; (d) 15 m.

Fig 13

3 With reference to *Fig 14*, the clockwise moment about A is:
(a) 70 Nm; (b) 10 Nm; (c) 60 Nm; (d) $5 \times R_B$ Nm

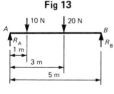

4 The force acting at B, (i.e. R_B) in *Fig 14* is:
(a) 16 N; (b) 20 N; (c) 15 N; (d) 14 N.

5 The force acting at A, (i.e. R_A) in *Fig 14* is:
(a) 16 N; (b) 10 N; (c) 15 N; (d) 14 N.

Fig 14

6 Which of the following statements is false
   for the beam shown in *Fig 15* if the beam
   is in equilibrium?
   (a) The anticlockwise moment is 27 N.
   (b) The force F is 9 N.
   (c) The reaction at the support, R is 18 N.
   (d) The beam cannot be in equilibrium for
       the given conditions.

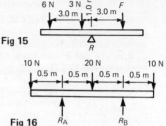

Fig 15

7 With reference to *Fig 16*, the reaction R_A is:
   (a) 10 N; (b) 30 N; (c) 20 N; (d) 40 N.

Fig 16

8 With reference to *Fig 16*, when moments are taken about point A, the sum of the he
   anticlockwise moments is:
   (a) 25 Nm; (b) 20 Nm; (c) 35 Nm; (d) 30 Nm.

9 With reference to *Fig 16*, when moments are taken about the right-hand end,
   the sum of the clockwise moments is:
   (a) 10 Nm; (b) 20 Nm; (c) 30 Nm; (d) 40 Nm.

10 With reference to *Fig 16*, which of the following statements is false?
    (a) $(5 + R_B) = 25$ Nm; (b) $R_A = R_B$;
    (c) $(10 \times 0.5) = (10 \times 1) + (10 \times 1.5) + R_A$; (d) $R_A + R_B = 40$ N

(c) CONVENTIONAL PROBLEMS

1 Determine the moment of a force of 25 N applied to a spanner at an effective
   length of 180 mm from the centre of a nut.   [4.5 Nm]

2 A moment of 7.5 Nm is required to turn a wheel. If a force of 37.5 N is applied to
   the rim of the wheel, calculate the effective distance from the rim to the hub of the
   wheel.   [200 mm]

3 Calculate the force required to produce a moment of 27 Nm on a shaft, when the
   effective distance from the centre of the shaft to the point of application of the
   force is 180 mm.   [150 N]

4 Determine distance d and the force acting at the support A for the force system
   shown in *Fig 17*, when the system is in equilibrium.   [50 mm; 3.8 kN]

5 If the 1 kN force shown in *Fig 17* is replaced by a force of F at a distance of
   250 mm to the left of R_A, find the value of F for the system to be in equilibrium.
                                                                            [560 N]

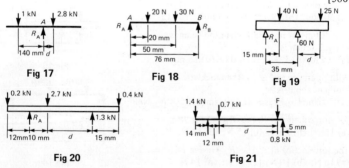

Fig 17

Fig 18

Fig 19

Fig 20

Fig 21

108

6 Determine the values of the forces acting at A and B for the force system shown in *Fig 18*. $[R_A = R_B = 25 \text{ N}]$

7 The forces acting on a beam are as shown in *Fig 19*. Neglecting the mass of the beam, find the value of R_A and distance d when the beam is in equilibrium.
$[5 \text{ N}; 25 \text{ mm}]$

8 Calculate the force R_A and distance d for the beam shown in *Fig 20*. The mass of the beam should be neglected and equilibrium conditions assumed.
$[2.0 \text{ kN}; 24 \text{ mm}]$

9 For the force system shown in *Fig 21*, find the values of F and d for the system to be in equilibrium.
$[1.0 \text{ kN}; 64 \text{ mm}]$

10 For the force system shown in *Fig 22*, determine distance d for forces R_A and R_B to be equal, assuming equilibrium conditions. $[80 \text{ m}]$

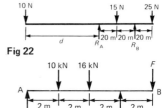

Fig 22

11 A simply supported beam AB is loaded as shown in *Fig 23*. Determine the load F in order that the reaction at A is zero.
$[36 \text{ kN}]$

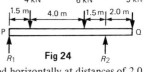

Fig 23

12 A uniform wooden beam, 4.8 m long, is supported at its left-hand end and also at 3.2 m from the left-hand end. The mass of the beam is equivalent to 200 N acting vertically downwards at its centre. Determine the reactions at the supports.
$[50 \text{ N}; 150 \text{ N}]$

13 For the simply supported beam PQ shown in *Fig 24*, determine (a) the reaction at each support; (b) the maximum force which can be applied at Q without losing equilibrium.
[(a) $R_1 = 3 \text{ kN}, R_2 = 12 \text{ kN}$; (b) 15.5 kN]

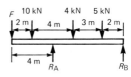

Fig 24

14 A uniform beam AB is 12 m long and is supported horizontally at distances of 2.0 m and 9.0 m from A. Loads of 60 kN, 104 kN, 50 kN and 40 kN act vertically downwards at A, 5.0 m from A, 7.0 m from A and at B respectively. Neglecting the mass of the beam, determine the reactions at the supports. $[133.7 \text{ kN}; 120.3 \text{ kN}]$

15 A uniform girder carrying point loads is shown in *Fig 25*. Determine the value of load F which causes the beam to just lift off the support B. $[3.25 \text{ kN}]$

Fig 25

9 Linear and angular motion

A. MAIN POINTS CONCERNING LINEAR AND ANGULAR MOTION

1 The unit of angular displacement is the **radian**,
where one radian is the angle subtended at
the centre of a circle by an arc equal in length
to the radius, see *Fig 1*. The relationship
between angle in radians (θ), arc length (s)
and radius of a circle (r) is:

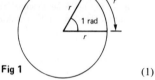

$$\boxed{s = r\theta} \quad \textbf{Fig 1}$$ (1)

Since the arc length of a complete circle is $2\pi r$ and the angle subtended at the centre
is $360°$, then from equation (1), for a complete circle, $2\pi r = r\theta$ or $\theta = 2\pi$ radians.

Thus, 2π radians corresponds to $360°$ (2)

2 (i) **Linear velocity** v is defined as the rate of change of linear displacement s with
respect to time t, and for motion in a straight line:

Linear velocity = $\dfrac{\text{change of displacement}}{\text{change of time}}$ i.e. $\boxed{v = \dfrac{s}{t}}$ (3)

The unit of linear velocity is metres per second (m/s).

(ii) **Angular velocity** The speed of revolution of a wheel or a shaft is usually
measured in revolutions per minute or revolutions per second but these units
do not form part of a coherent system of units. The basis used in S.I. units is the
angle turned through in one second.

Angular velocity is defined as the rate of change of angular displacement θ,
with respect to time t, and for an object rotating about a fixed axis at a constant
speed:

Angular velocity = $\dfrac{\text{angle turned through}}{\text{time taken}}$, i.e. $\boxed{\omega = \dfrac{\theta}{t}}$ (4)

The unit of angular velocity is radians per second (rad/s).

An object rotating at a constant speed of n revolutions per second subtends
an angle of $2\pi n$ radians in one second, that is, its angular velocity (5)

$$\boxed{\omega = 2\pi n \text{ rad/s}}$$

(iii) From equation (1), $s = r\theta$ and from equation (4), $\theta = \omega t$, hence $s = r\omega t$,

or $\frac{s}{t} = \omega r$. However, from equation (3), $v = \frac{s}{t}$, hence $\boxed{v = \omega r}$ (6)

Equation (6) gives the relationship between linear velocity, v, and angular velocity, ω.

3 (i) **Linear acceleration**, a, is defined as the rate of change of linear velocity with respect to time. For an object whose linear velocity is increasing uniformly:

$$\text{linear acceleration} = \frac{\text{change of linear velocity}}{\text{time taken}}$$

i.e. $\boxed{a = \dfrac{v_2 - v_1}{t}}$ (7)

The unit of linear acceleration is metres per second squared (m/s^2).

Rewriting equation (7) with v_2 as the subject of the formula gives:

$$\boxed{v_2 = v_1 + at}$$ (8)

(ii) **Angular acceleration** α, is defined as the rate of change of angular velocity with respect to time. For an object whose angular velocity is increasing uniformly:

$$\text{Angular acceleration} = \frac{\text{change of angular velocity}}{\text{time taken}}$$

that is, $\boxed{\alpha = \dfrac{\omega_2 - \omega_1}{t}}$ (9)

The unit of angular acceleration is radians per second squared (rad/s^2).

Rewriting equation (9) with ω_2 as the subject of the formula gives:

$$\boxed{\omega_2 = \omega_1 + \alpha t}$$ (10)

(iii) From equation (6), $v = \omega r$. For motion in a circle having a constant radius r, $v_2 = \omega_2 r$ and $v_1 = \omega_1 r$, hence equation (7) can be rewritten as:

$$a = \frac{\omega_2 r - \omega_1 r}{t} = \frac{r(\omega_2 - \omega_1)}{t}$$

But from equation (9), $\dfrac{\omega_2 - \omega_1}{t} = \alpha$

Hence $\boxed{a = r\alpha}$ (11)

Equation (11) gives the relationship between linear acceleration a and angular acceleration α.

4 (i) From equation (3), $s = vt$, and if the linear velocity is changing uniformly from v_1 to v_2, then $s = $ mean linear velocity $\times$ time,

i.e. $\boxed{s = \left(\dfrac{v_1 + v_2}{2}\right) t}$ (12)

(ii) From equation (4), $\theta = \omega t$, and if the angular velocity is changing uniformly from ω_1 to ω_2, then $\theta = $ mean angular velocity $\times$ time,

i.e. $\boxed{\theta = \left(\dfrac{\omega_1 + \omega_2}{2}\right) t}$ (13)

(iii) Two further equations of linear motion may be derived from equations (8) and (11):

$$\boxed{s = v_1 t + \tfrac{1}{2} at^2}$$ (14)

111

and

$$v_2{}^2 = v_1{}^2 + 2as \tag{15}$$

(iv) Two further equations of angular motion may be derived from equations (10) and (12):

$$\theta = \omega_1 t + \tfrac{1}{2}\alpha t^2 \tag{15}$$

and

$$\omega_2{}^2 = \omega_1{}^2 + 2\alpha\theta \tag{16}$$

5 Table 1 summarises the principal equations of linear and angular motion for uniform changes in velocities and constant accelerations and also gives the relationship between linear and angular quantities.

TABLE 1

s = arc length (m)	r = radius of circle (m)
t = time (s)	θ = angle (rad)
v = linear velocity (m/s)	ω = angular velocity (rad/s)
v_1 = initial linear velocity (m/s)	ω_1 = initial angular velocity (rad/s)
v_2 = final linear velocity (m/s)	ω_2 = final angular velocity (rad/s)
a = linear acceleration (m/s²)	α = angular acceleration (rad/s²)
n = speed of revolutions (revolutions per second)	

Equation No.	Linear motion	Angular motion
1	$s = r\theta$ m	
2		2π rad $= 360°$
3 and 4	$v = \dfrac{s}{t}$ m/s	$\omega = \dfrac{\theta}{t}$ rad/s
5		$\omega = 2\pi n$ rad/s
6	$v = \omega r$ m/s	
8 and 10	$v_2 = (v_1 + at)$ m/s	$\omega_2 = (\omega_1 + \alpha t)$ rad/s
11	$a = r\alpha$ m/s²	
12 and 13	$s = \left(\dfrac{v_1 + v_2}{2}\right) t$ m	$\theta = \left(\dfrac{\omega_1 + \omega_2}{2}\right) t$ rad
14 and 16	$s = (v_1 t + \tfrac{1}{2}at^2)$ m	$\theta = (\omega_1 t + \tfrac{1}{2}\alpha t^2)$ rad
15 and 17	$v_2{}^2 = (v_1{}^2 + 2as)$ (m/s)²	$\omega_2{}^2 = (\omega_1{}^2 + 2\alpha\theta)$ (rad/s)²

6 Quantities used in engineering and science can be divided into two groups:
   (a) **Scalar quantities** have a size or magnitude only and need no other information to specify them. Thus 10 centimetres, 50 seconds, 7 litres and 3 kilograms are all examples of scalar quantities.
   (b) **Vector quantities** have both a size or magnitude, and a direction, called the line of action of the quantity. Thus, a velocity of 50 kilometres per hour due east, and an acceleration of 9.8 metres per second squared acting vertically downwards, are both vector quantities.

7 A vector quantity is represented by a straight line lying along the line of action of the quantity and having a length which is proportional to the size of the quantity. Thus **ab** in *Fig 2* represents a velocity of 20 m/s, whose line of action is due

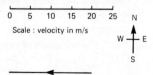

Fig 2

west. The bold letters, **ab**, indicate a
vector quantity and the order of the
letters indicate that the time of action
is from a to **b**.

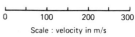

Scale : velocity in m/s

8   Consider two aircraft A and B flying
at a constant altitude, A travelling due
north at 200 m/s and B travelling 30°
east of north, written N 30° E, at
300 m/s, as shown in *Fig 3*. Relative to a fixed point o, **oa** represents
the velocity of A and **ob** the velocity
of B. The velocity of B relative to A,
that is the velocity at which B seems
to be travelling to an observer on A,
is given by **ab**, and by measurement
is 160 m/s in a direction E 22° N.

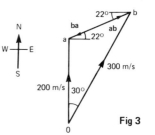

Fig 3

The velocity of A relative to B, that is, the velocity at which A seems to be travelling
to an observer on B, is given by **ba** and by measurement is 160 m/s in a direction
W 22° S.

B. WORKED PROBLEMS ON LINEAR AND ANGULAR MOTION

Problem 1 A wheel of diameter 540 mm is rotating at $\frac{1500}{\pi}$ rev/min. Calculate
the angular velocity of the wheel and the linear velocity of a point on the rim of
the wheel.

From equation 5, angular velocity $\omega = 2\pi n$, where n is the speed of revolution in
revolutions per second, i.e. $n = \frac{1500}{60\pi}$ revolutions per second.

Thus $\omega = 2\pi \frac{1500}{60\pi} = \mathbf{50\ rad/s}$

The linear velocity of a point on the rim, $v = \omega r$, where r is the radius of the wheel,
i.e. $\frac{0.54}{2}$ or 0.27 m.

Thus $v = 50 \times 0.27 = \mathbf{13.5\ m/s}$

Problem 2 A car is travelling at 64.8 km/h and has wheels of diameter 600 mm.
(a) Find the angular velocity of the wheels in both rad/s and revolutions per minute.
(b) If the speed remains constant for 1.44 km, determine the number of revolutions
made by a wheel, assuming no slipping occurs.

(a) 64.8 km/h = $64.8 \frac{km}{h} \times 1000 \frac{m}{km} \times \frac{1}{3600} \frac{h}{s}$ = 18 m/s

That is, the linear velocity, v, is 18 m/s.

The radius of a wheel is $\frac{600}{2}$ mm = 0.3 m.

From equation 6, $v = \omega r$, hence $\omega = \frac{v}{r}$

That is, the angular velocity, $\omega = \frac{18}{0.3} = \mathbf{60\ rad/s}$.

From equation 5, $\omega = 2\pi n$, where n is in revolutions per second.

Hence $n = \dfrac{\omega}{2\pi}$ and angular speed of a wheel in revolutions per minute is $\dfrac{60\omega}{2\pi}$.

But $\omega = 60$ rad/s, hence angular speed $= \dfrac{60 \times 60}{2\pi} = 573$ **revolutions per minute.**

(b) From equation 3, time taken to travel 1.44 km at a constant speed of 18 m/s is

$\dfrac{1440 \text{ m}}{18 \text{ m/s}} = 80$ s.

Since a wheel is rotating at 573 revolutions per minute, then in $\dfrac{80}{60}$ minutes it

makes $\dfrac{573 \times 80}{60}$, that is, **764 revolutions**

Problem 3 The speed of a shaft increases uniformly from 300 revolutions per minute to 800 revolutions per minute in 10 s. Find the angular acceleration, correct to four significant figures.

From equation 10, $\omega_2 = \omega_1 + \alpha t$, hence $\alpha = \dfrac{\omega_2 - \omega_1}{t}$

Initial angular velocity $\omega_1 = 300$ revolutions per minute

$= \dfrac{300}{60}$ revolutions per second $= \dfrac{300 \times 2\pi}{60}$ rad/s

Final angular velocity, $\omega_2 = \dfrac{800 \times 2\pi}{60}$ rad/s

Hence, angular acceleration $\alpha = \left(\dfrac{800 \times 2\pi}{60} - \dfrac{300 \times 2\pi}{60}\right) \Big/ 10$ rad/s^2

$= \dfrac{500 \times 2\pi}{60 \times 10} = 5.236$ rad/s^2

Problem 4 If the diameter of the shaft in *Problem 3* is 50 mm, determine the linear acceleration of the shaft, correct to three significant figures.

From equation 11, $a = r\alpha$

The shaft radius is $\dfrac{0.05}{2}$ m and the angular acceleration α is 5.236 rad/s^2, thus the

linear acceleration, $a = \dfrac{0.05}{2} \times 5.236 = 0.131$ m/s^2

Problem 5 Find the number of revolutions made by the shaft in *Problem 3* during the 10 s it is accelerating.

From equation 13, angle turned through $\theta = \left(\dfrac{\omega_1 + \omega_2}{2}\right)t = \tfrac{1}{2}(\omega_1 + \omega_2)t$

$= \tfrac{1}{2}\left(\dfrac{300 \times 2\pi}{60} + \dfrac{800 \times 2\pi}{60}\right)(10)$ rad

But there are 2π rad in 1 revolution.

Hence, number of revolutions $= \tfrac{1}{2}\left(\dfrac{300 \times 2\pi}{60} + \dfrac{800 \times 2\pi}{60}\right)\left(\dfrac{10}{2\pi}\right) = \tfrac{1}{2}\left(\dfrac{1100}{60}\right)(10)$

$= \dfrac{1100}{60} \times 5 = 91\tfrac{2}{3}$ **revolutions**

Problem 6 The shaft of an electric motor, initially at rest, accelerates uniformly for 0.4 s at 15 rad/s². Determine the angle turned through by the shaft, in radians, in this time.

From equation 16, $\theta = \omega_1 t + \frac{1}{2}\alpha t^2$

Since the shaft is initially at rest, $\omega_1 = 0$ and $\theta = \frac{1}{2}\alpha t^2$

The angular acceleration, α is 15 rad/s² and time t is 0.4 s, hence

angle turned through, $\theta = \frac{1}{2} \times 15 \times 0.4^2 = 1.2$ **rad**

Problem 7 A flywheel accelerates uniformly at 2.05 rad/s² until it is rotating at 1500 revolutions per minute. If it completes 5 revolutions during the time it is accelerating, determine its initial angular velocity in rad/s, correct to four significant figures.

Since the final angular velocity is 1500 revolutions per minute,

$\omega_2 = 1500 \dfrac{\text{rev}}{\text{min}} \times \dfrac{1 \text{ min}}{60 \text{ s}} \times \dfrac{2\pi \text{ rad}}{1 \text{ rev}} = 50\pi$ rad/s

5 revolutions = 5 rev $\times \dfrac{2\pi \text{ rad}}{1 \text{ rev}} = 10\pi$ rad

From equation 17, $\omega_2{}^2 = (\omega_1{}^2 + 2\alpha\theta)(\text{rad/s})^2$

i.e. $(50\pi)^2 = \omega_1{}^2 + 2 \times 2.05 \times 10\pi$

$\quad \omega_1{}^2 = (50\pi)^2 - (2 \times 2.05 \times 10\pi)$

$\quad\quad = (50\pi)^2 - 41\pi = 24\,545$

i.e. $\omega_1 = 156.7$ rad/s

Thus the initial angular velocity is 156.7 rad/s, correct to four significant figures

Problem 8 Two cars are travelling on horizontal roads in straight lines, car A at 70 km/h at N 10° E and car B at 50 km/h at W 60° N. Determine, by drawing a vector diagram to scale, the velocity of car A relative to car B.

With reference to *Fig 4(a)*, **oa** represents the velocity of car A relative to a fixed point O, and *ob* represents the velocity of car B relative to a fixed point O. The velocity of car A relative to car B is given by **ba** and by measurement is **45 km/h in a direction of E 35° N.**

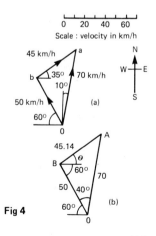

Fig 4

Problem 9 Verify the result obtained in *Problem 8* by calculation.

The triangle shown in *Fig 4(b)* is similar to the vector diagram shown in *Fig 4(a)*. Angle BOA is $180° - (60° + 80°)$, that is, $40°$. Using the cosine rule:

$$BA^2 = 50^2 + 70^2 - 2 \times 50 \times 70 \times \cos 40°$$

i.e. **BA = 45.14**

Using the sine rule, $\dfrac{50}{\sin \angle BAO} = \dfrac{45.14}{\sin 40°}$

$$\sin \angle BAO = \frac{50 \sin 40°}{45.14} = 0.7120$$

Hence angle   BAO = $45° \; 24'$

Thus angle   ABO = $180 - (40 + 45° \; 24') = 94° \; 36'$

and angle $\theta = 94° \; 36' - 60° = 34° \; 36'$

Thus **ba is 45.14 km/h in a direction E 34° 36′ N by calculation.**

Problem 10 A crane is moving in a straight line with a constant horizontal speed of 2 m/s. At the same time it is lifting a load at a vertical speed of 5 m/s. Calculate the velocity of the load relative to a fixed point on the earth's surface.

A vector diagram depicting the motion of the crane and load is shown in *Fig 5*. oa represents the velocity of the crane relative to a fixed point on the earth's surface and ab represents the velocity of the load relative to the crane. The velocity of the load relative to the fixed point on the earth's surface is **ob**.

By Pythagoras' theorem: $ob^2 = oa^2 + ab^2$

$$= 4 + 25 = 29$$

Hence ob $= \sqrt{29} = 5.385$ m/s

Tan $\theta = \dfrac{5}{2} = 2.5$

Hence $\theta = \arctan 2.5 = 68° \; 12'$

That is, the velocity of the load relative to a fixed point on the earth's surface is **5.385 m/s in a direction 68° 12′ to the motion of the crane**

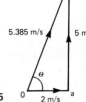

Fig 5

C. FURTHER PROBLEMS ON LINEAR AND ANGULAR MOTION

(a) SHORT ANSWER PROBLEMS

1   State and define the unit of angular displacement.

2   Write down the formula connecting an angle, arc length and the radius of a circle.

3   Define linear velocity and state its unit.

4   Define angular velocity and state its unit.

5   Write down a formula connecting angular velocity and revolutions per second in coherent units.

6   State the formula connecting linear and angular velocity.

7   Define linear acceleration and state its unit.

8   Define angular acceleration and state its unit.

9   Write down the formula connecting linear and angular acceleration.

10 Define a scalar quantity and give two examples.

11 Define a vector quantity and give two examples.

(b) MULTI-CHOICE PROBLEMS (answers on page 172)

1   An angle of 2 rad at the centre of a circle subtends an arc length of 40 mm at the
    circumference of the circle. The radius of the circle is:

    (a) 40π mm; (b) 80 mm; (c) 20 mm; (d) $\dfrac{40}{\pi}$ mm.

2   A point on a wheel has a constant angular velocity of 3 rad/s. The angle turned
    through in 15 seconds is:
    (a) 45 rad; (b) 10π rad; (c) 5 rad; (d) 90π rad.

3   An angular velocity of 60 revolutions per minute is the same as:

    (a) $\dfrac{1}{2\pi}$ rad/s; (b) 120π rad/s; (c) $\dfrac{30}{\pi}$ rad/s; (d) 2π rad/s.

4   A wheel of radius 15 mm has an angular velocity of 10 rad/s. A point on the rim
    of the wheel has a linear velocity of:
    (a) 300π mm/s; (b) $\frac{2}{3}$ mm/s; (c) 150 mm/s; (d) 1.5 mm/s.

5   The shaft of an electric motor is rotating at 20 rad/s and its speed is increased
    uniformly to 40 rad/s in 5 s. The angular acceleration of the shaft is:
    (a) 4000 rad/s²; (b) 4 rad/s²; (c) 160 rad/s²; (d) 12 rad/s²

6   A point on a flywheel of radius 0.5 m has a uniform linear acceleration of 2 m/s².
    Its angular acceleration is:
    (a) 2.5 rad/s²; (b) 0.25 rad/s²; (c) 1 rad/s²; (d) 4 rad/s².

A car accelerates uniformly from 10 m/s to 20 m/s over a distance of 150 m. The
wheels of the car each have a radius of 250 mm.
In *Problems 7 to 10* use this data to determine the quantities stated, selecting the
correct answer from those given below.
(a) 2.5 rad/s; (b) $\frac{1}{5}$ s; (c) 1 m/s²; (d) 20 rad/s; (e) 1 rad/s²; (f) $\frac{1}{4}$ rad/s²; (g) 10 s;
(h) 100 m/s²; (i) 40 rad/s; (j) 3 m/s²; (k) 4 rad/s²; (l) 5 s.
7   The time the car is accelerating.
8   The initial angular velocity of each of the wheels.
9   The linear acceleration of a point on each of the wheels.
10 The angular acceleration of each of the wheels.

(c) CONVENTIONAL PROBLEMS

1   A pulley driving a belt has a diameter of 360 mm and is turning at $2700/\pi$ revolutions
    per minute. Find the angular velocity of the pulley and the linear velocity of the
    belt assuming that no slip occurs.   [$\omega = 90$ rad/s; $v = 16.2$ m/s]

2   A bicycle is travelling at 36 km/h and the diameter of the wheels of the bicycle is
    500 mm. Determine the angular velocity of the wheels of the bicycle and the linear
    velocity of a point on the rim of one of the wheels.   [$\omega = 40$ rad/s; $v = 10$ m/s]

3 A flywheel rotating with an angular velocity of 200 rad/s is uniformly accelerated at a rate of 5 rad/s² for 15 s. Find the final angular velocity of the flywheel both in rad/s and revolutions per minute. [275 rad/s; 8250/π rev/min]

4 A disc accelerates uniformly from 300 revolutions per minute to 600 revolutions per minute in 25 s. Determine its angular acceleration and the linear acceleration of a point on the rim of the disc, if the radius of the disc is 250 mm.

[0.4π rad/s² ; 0.1π m/s²]

5 Calculate the number of revolutions the disc in *Problem 4* makes during the 25 s accelerating period. [187.5 revolutions]

6 A grinding wheel makes 300 revolutions when slowing down uniformly from 1000 rad/s to 400 rad/s. Find the time for this reduction in speed. [2.693 s]

7 Find the angular retardation for the grinding wheel in *Problem 6*. [222.8 rad/s²]

8 A pulley is accelerated uniformly from rest at a rate of 8 rad/s². After 20 s the acceleration stops and the pulley runs at constant speed for 2 min, and then the pulley comes uniformly to rest after a further 40 s. Calculate:
(a) the angular velocity after the period of acceleration;
(b) the deceleration;
(c) the total number of revolutions made by the pulley.

[(a) 160 rad/s; (b) 4 rad/s² ; (c) 12 000/π rev]

9 A ship is sailing due east with a uniform speed of 7.5 m/s relative to the sea. If the tide has a velocity 2 m/s in a north-westerly direction, find the velocity of the ship relative to the sea bed. [6.248 m/s at E 13° 5′ N]

10 A lorry is moving along a straight road at a constant speed of 54 km/h. The tip of its windscreen wiper blade has a linear velocity, when in a vertical position of 4 m/s. Find the velocity of the tip of the wiper blade relative to the road when in this vertical position. [15.52 m/s at 14° 56′]

11 A fork-lift truck is moving in a straight line at a constant speed of 5 m/s and at the same time, a pallet is being lowered at a constant speed of 2 m/s. Determine the velocity of the pallet relative to the earth. [5.385 m/s at (−21° 48′)]

10 Force, mass and acceleration

A. MAIN POINTS CONCERNING FORCE, MASS AND ACCELERATION

1 To make a stationary object move or to change the direction in which the object is moving requires a force to be applied externally to the object. This concept is known as **Newton's first law of motion** and may be stated as:

'An object remains in a state of rest, or continues in a state of uniform motion in a straight line, unless it is acted on by an externally applied force '

2 Since a force is necessary to produce a change of motion, an object must have some resistance to a change in its motion. The force necessary to give a stationary pram a given acceleration is far less than the force necessary to give a stationary car the same acceleration. The resistance to a change in motion is called the **inertia** of an object and the amount of inertia depends on the mass of the object. Since a car has a much larger mass than a pram, the inertia of a car is much larger than that of a pram.

3 **Newton's second law of motion** may be stated as:

'The acceleration of an object acted upon by an external force is proportional to the force and is in the same direction as the force.'

Thus, force ∝ acceleration, or force = a constant × acceleration, this constant of proportionality being the mass of the object, i.e.

force = mass × acceleration

The unit of force is the newton (N) and is defined in terms of mass and acceleration. One newton is the force required to give a mass of 1 kilogram an acceleration of 1 metre per second squared. Thus

$F = ma$, where

F is the force in newtons (N), m is the mass in kilograms (kg) and a is the acceleration in metres per second squared (m/s²), i.e.,

$$1 \text{ N} = 1 \frac{\text{kg m}}{\text{s}^2}$$

(It follows that $\frac{1\text{m}}{\text{s}^2} = \frac{1\text{N}}{\text{kg}}$. Hence a gravitational acceleration of 9.81 m/s² is the same as a gravitational field of 9.81 N/kg)

4 **Newton's third law of motion** may be stated as:

'For every force, there is an equal and opposite reacting force.'

Thus, an object on, say a table, exerts a downward force on the table and the table exerts an equal upward force on the object, known as **a reaction force** or just a **reaction**.

119

5   When an object moves in a circular path at
    constant speed, its direction of motion is con-
    tinually changing and hence its velocity
    (which depends on both magnitude **and**
    direction) is also continually changing. Since
    acceleration is the

    $$\frac{\text{change in velocity}}{\text{time taken}},$$

    the object has an acceleration.

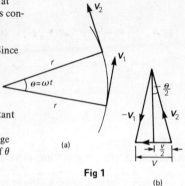

    Let the object be moving with a constant
angular velocity of ω and a tangential
velocity of magnitude v and let the change
of velocity for a small change of angle of θ
$(= \omega t)$ be V.

Fig 1

Then, $v_2 - v_1 = V$

The vector diagram is shown in *Fig 1(b)* and since the magnitudes of v_1 and v_2
are the same, i.e. v, the vector diagram is also an isosceles triangle.

Bisecting the angle between v_2 and v_1 gives: $\sin\frac{\theta}{2} = \frac{V/2}{v_2} = \frac{V}{2v}$

i.e. $V = 2v \sin\frac{\theta}{2}$ (1)

Since $\theta = \omega t$, $t = \frac{\theta}{\omega}$ (2)

Dividing (1) by (2) gives:

$$\frac{V}{t} = \frac{2v \sin\frac{\theta}{2}}{\frac{\theta}{\omega}} = \frac{v\omega \sin\frac{\theta}{2}}{\frac{\theta}{2}}$$

For small angles, $\dfrac{\sin\frac{\theta}{2}}{\frac{\theta}{2}}$ is very nearly equal to unity.

Hence, $\dfrac{V}{t} = \dfrac{\text{change of velocity}}{\text{change of time}} = \text{acceleration}, \ a = v\omega$

But, $\omega = \dfrac{v}{r}$, thus $v\omega = v \cdot \dfrac{v}{r} = \dfrac{v^2}{r}$

That is, the acceleration a is $\dfrac{v^2}{r}$ and is towards the centre of the circle of motion
(along V). It is called the **centripetal acceleration**. If the mass of the rotating object
is m, then by Newton's second law, the **centripetal force** is mv^2/r, and its direction
is towards the centre of the circle of motion.

B. WORKED PROBLEMS ON FORCE, MASS AND ACCELERATION

Problem 1 Calculate the force needed to accelerate a boat of mass 20 t uniformly
from rest to a speed of 21.6 km/h in 10 minutes.

The mass of the boat, m, is 20 t, that is 20 000 kg. The law of motion, $v_2 = v_1 + at$
can be used to determine the acceleration a. The initial velocity v_1 is zero.

The final velocity v_2 is 21.6 km/h, that is $\dfrac{21.6}{3.6}$ or 6 m/s. The time t is 10 min,
that is 600 s.

120

Thus, $6 = 0 + a \times 600$ or $a = \dfrac{6}{600} = 0.01$ m/s^2

From Newton's second law, $F = ma$

i.e. Force $= 20\,000 \times 0.01$ N

$\qquad\qquad = \mathbf{200\ N}$

Problem 2 The moving head of a machine tool requires a force of 1.2 N to bring it to rest in 0.8 s from a cutting speed of 30 m/min. Find the mass of the moving head.

From Newton's second law, $F = ma$, thus $m = \dfrac{F}{a}$, where force is given as 1.2 N.

The law of motion $v_2 = v_1 + at$ can be used to find acceleration a, where $v_2 = 0$, $v_1 = 30$ m/min, that is $\dfrac{30}{60}$ or 0.5 m/s, and $t = 0.8$ s. Thus, $0 = 0.5 + a \times 0.8$,

i.e., $a = -\dfrac{0.5}{0.8} = -0.625$ m/s^2, or a retardation of 0.625 m/s^2.

Thus the mass, $m = \dfrac{1.2}{0.625} = \mathbf{1.92\ kg}$

Problem 3 A lorry of mass 1350 kg accelerates uniformly from 9 km/h to reach a speed of 45 km/h in 18 s. Determine (a) the acceleration of the lorry; (b) the uniform force needed to accelerate the lorry.

(a) The law of motion $v_2 = v_1 + at$ can be used to determine the acceleration, where final velocity $v_2 = \dfrac{45}{3.6}$ m/s, initial velocity $v_1 = \dfrac{9}{3.6}$ m/s and time t is 18 s.

Thus, $\dfrac{45}{3.6} = \dfrac{9}{3.6} + a \cdot 18$

$a = \dfrac{1}{18}\left(\dfrac{45}{3.6} - \dfrac{9}{3.6}\right) = \dfrac{1}{18} \times \dfrac{36}{3.6} = \dfrac{10}{18} = \mathbf{\dfrac{5}{9}}$ m/s^2

(b) From Newton's second law of motion, $F = ma = 1350 \times \dfrac{5}{9} = \mathbf{750\ N}$

Problem 4 Find the weight of an object of mass 1.6 kg at a point on the earth's surface where the gravitational field is 9.81 N/kg.

The weight of an object is the force acting vertically downwards due to the force of gravity acting on the object. Thus:

Weight $=$ force acting vertically downwards $=$ mass $\times$ gravitational field

$\qquad\quad = 1.6 \times 9.81 = \mathbf{15.696\ N}$

Problem 5 A bucket of cement of mass 40kg is tied to the end of a rope connected to a hoist. Calculate the tension in the rope when the bucket is suspended but stationary. Take the gravitational field, g as 9.81 N/kg

The **tension in** the rope is the same as the force acting in the rope. The force acting vertically downwards due to the weight of the bucket must be equal to the force acting upwards in the rope, i.e. the tension.

Weight of bucket of cement, $F = mg = 40 \times 9.81 = 392.4$ N

Thus, the tension in the rope is also **392.4 N**

Problem 6 The bucket of cement in *Problem 5* is now hoisted vertically upwards with a uniform acceleration of 0.4 m/s². Calculate the tension in the rope during the period of acceleration.

With reference to *Fig 2(a)*, the forces acting on the bucket are:
(i) a tension (or force) of T acting in the rope;
(ii) a force of mg acting vertically downwards i.e. the weight of the bucket and cement;

Fig 2

The resultant force $F = T - mg$
Hence, from para 3, $ma = T - mg$
$40 \times 0.4 = T - 40 \times 9.81$
giving $T = 408.4$ N
By comparing this result with that of *Problem 5*, it can be seen that there is an increase in the tension in the rope when an object is accelerating upwards.

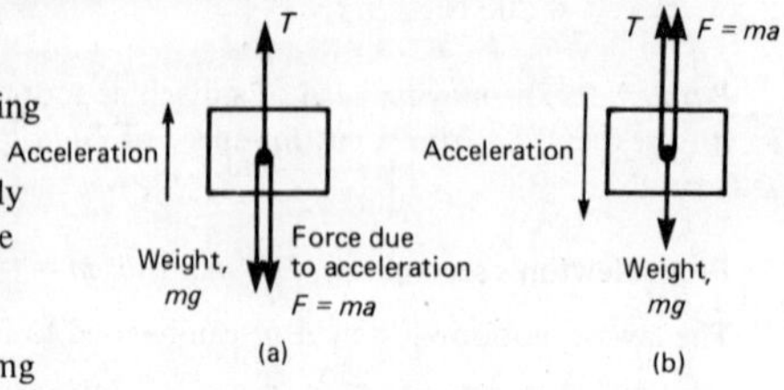

Problem 7 The bucket of cement in *Problem 5* is now lowered vertically downwards with a uniform acceleration of 1.4 m/s². Calculate the tension in the rope during the period of acceleration.

With reference to *Fig 2(b)*, the forces acting on the bucket are:
(i) a tension (or force) of T acting vertically upwards;
(ii) A force of mg acting vertically downwards i.e. the weight of the bucket and cement;

The resultant force $F = mg - T$. Hence, from para 3, $ma = mg - T$

i.e. $T = m(g-a) = 40(9.81 - 1.4) = \mathbf{336.4 \ N}$

By comparing this result with that of *Problem 5*, it can be seen that there is a decrease in the tension in the rope when an object is accelerating downwards.

Problem 8 Two masses are suspended vertically by a thread over a pulley and are at the same height. One of the masses is 10 kg and the system has an acceleration of 1.2 m/s² when released, the acceleration being towards the 10 kg mass. Determine the value of the other mass, assuming g is 9.81 m/s² and all losses and the mass of the pulley are neglected.

Let the unknown mass be m, the resultant force be F, and the tension in the thread be T (see *Fig 3*). Since the 10 kg mass accelerates downwards, the tension T in the thread is less than mg,

Hence, resultant force $= 10g - T$
and the acceleration of 10 kg mass $= \dfrac{10g - T}{10} = 1.2$ \qquad (1)
Resultant force on mass, m $= T - mg$

and acceleration of mass, $m = \dfrac{T - mg}{m} = 1.2$ (2)

Fig 3

From equation (1), $10g - T = 12$
From equation (2), $T - mg = 1.2\,m$
Adding gives: $10g - mg = 12 + 1.2\,m$
$98.1 - 12 = m(1.2 + 9.81)$
i.e. $m = \dfrac{86.1}{11.01} = 7.82$ kg.

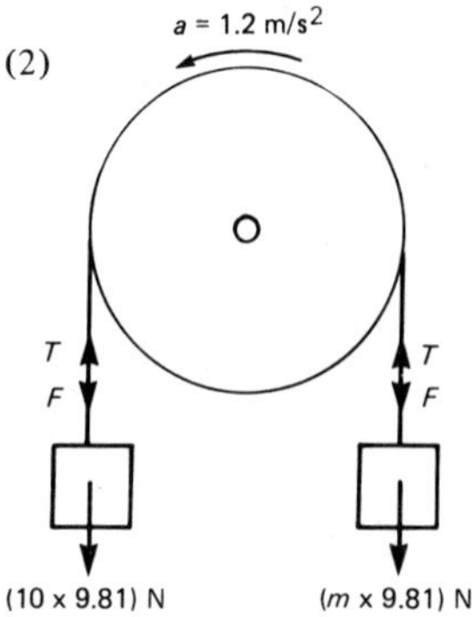

Problem 9 A vehicle of mass 750 kg travels round a bend of radius 150 m, at 50.4 km/h. Determine the centripetal force acting on the vehicle.

From para. 5, the centripetal force is given by $\dfrac{mv^2}{r}$ and its direction is towards the centre of the circle.

$m = 750$ kg, $v = 50.4$ km/h $= \dfrac{50.4}{3.6}$ m/s $= 14$ m/s, $r = 150$ m.

Thus, centripetal force $= \dfrac{750 \times (14)^2}{150} = \mathbf{980\ N}$

Problem 10 An object is suspended by a thread 250 mm long and both object and thread move in a horizontal circle with a constant angular velocity of 2.0 rad/s. If the tension in the thread is 12.5 N, determine the mass of the object.

From para. 5, centripetal force (i.e. tension in thread) $= \dfrac{mv^2}{r} = 12.5$ N.

The angular velocity $\omega = 2.0$ rad/s and radius $r = 250$ mm $= 0.25$ m.
Since linear velocity $v = \omega r$, $v = 2.0 \times 0.25 = 0.5$ m/s

Since $F = \dfrac{mv^2}{r}$, then $m = \dfrac{Fr}{v^2}$

i.e. mass of object, $m = \dfrac{12.5 \times 0.25}{0.5^2} = \mathbf{12.5\ kg}$

Problem 11 An aircraft is turning at constant altitude, the turn following the arc of a circle of radius 1.5 km. If the maximum allowable acceleration of the aircraft is 2.5 g, determine the maximum speed of the turn in km/h. Take g as 9.8 m/s².

From para. 5, the acceleration of an object turning in a circle is $\dfrac{v^2}{r}$.
Thus, to determine the maximum speed of turn
$\dfrac{v^2}{r} = 2.5\,g$
$v = \sqrt{(2.5\ gr)} = \sqrt{(2.5 \times 9.8 \times 1500)} = \sqrt{36\ 750} = 191.7$ m/s
191.7 m/s $= 191.7 \times 3.6$ km/h $= \mathbf{690\ km/h}$

C. FURTHER PROBLEMS ON FORCE, MASS AND ACCELERATION

(a) SHORT ANSWER PROBLEMS

1   State Newton's first law of motion.

2   Describe what is meant by the inertia of an object.

3   State Newton's second law of motion.

4   Define the newton.

5   State Newton's third law of motion.

6   Explain why an object moving round a circle at a constant angular velocity has an acceleration.

7   Define centripetal acceleration in symbols.

8   Define centripetal force in symbols.

(b) MULTI-CHOICE PROBLEMS (answers on page 173)

A man of mass 75 kg is standing in a lift of mass 500 kg. Use this data to determine the answers to *Problems 1 to 6*. Take g as 10 m/s².

1   The tension in a cable when the lift is moving at a constant speed vertically upward is
    (a) 4250 N; (b) 5750 N; (c) 4600 N; (d) 6900 N

2   The tension in the cable supporting the lift when the lift is moving at a constant speed vertically downwards is:
    (a) 4250 N; (b) 5750 N; (c) 4600 N; (d) 6900 N

3   The reaction force between the man and the floor of the lift when the lift is travelling at a constant speed vertically upwards is:
    (a) 750 N; (b) 900 N; (c) 600 N; (d) 475 N

4   The reaction force between the man and the floor of the lift when the lift is travelling at a constant speed vertically downwards is:
    (a) 750 N; (b) 900 N; (c) 600 N; (d) 475 N.

5   The reaction force between the man and the floor of the lift when the lift is accelerating at a rate of 2 m/s² vertically upwards is:
    (a) 750 N; (b) 900 N; (c) 600 N; (d) 475 N.

6   The reaction force between the man and the floor of the lift when the lift is accelerating at a rate of 2 m/s² vertically downwards is:
    (a) 750 N; (b) 900 N; (c) 600 N; (d) 475 N.

A ball of mass 0.5 kg is tied to a thread and rotated at a constant angular velocity of 10 rad/s in a circle of radius 1 m. Use this data to determine the answers to *Problems 7 and 8*.

7   The centripetal acceleration is:

    (a) 50 m/s²; (b) $\frac{100}{2\pi}$ m/s²; (c) $\frac{50}{2\pi}$ m/s²; (d) 100 m/s².

8   The tension in the thread is:

    (a) 25 N; (b) $\frac{50}{2\pi}$ N; (c) $\frac{25}{2\pi}$ N; (d) 50 N.

9 Which of the following statements is false?
   (a) An externally applied force is needed to change the direction of a moving object.
   (b) For every force, there is an equal and opposite reaction force.
   (c) A body travelling at a constant velocity in a circle has no acceleration.
   (d) Centripetal acceleration acts towards the centre of the circle of motion.

10 Which of the following statements is true?
   (a) The acceleration of an object acted upon by an external force is proportional to the force and acts in the opposite direction to the force.
   (b) The inertia of an object is the resistance of an object to a change in motion.
   (c) The tension in a cable supporting a lift is greater when the lift is moving with a constant speed vertically upwards than when it is stationary.
   (d) The tension in a cable supporting a lift remains constant irrespective of its motion

(c) CONVENTIONAL PROBLEMS

(Take g as 9.81 m/s^2, and express answers to three significant figure accuracy.)

1 A car initially at rest accelerates uniformly to a speed of 55 km/h in 14 s. Determine the accelerating force required if the mass of the car is 800 kg.   [873 N]

2 The brakes are applied on the car in *Problem 1* when travelling at 55 km/h and it comes to rest uniformly in a distance of 50 m. Calculate the braking force and the time for the car to come to rest.   [1.87 kN; 6.55 s]

3 The tension in a rope lifting a crate vertically upwards is 2.8 kN. Determine its acceleration if the mass of the crate is 270 kg.   [0.560 m/s^2]

4 A ship is travelling at 18 km/h when it stops its engines. It drifts for a distance of 0.6 km and its speed is then 14 km/h. Determine the value of the forces opposing the motion of the ship, assuming the reduction in speed is uniform and the mass of the ship is 2000 t.   [16.5 kN]

A cage having a mass of 2 t is being lowered down a mine shaft. It moves from rest with an acceleration of 4 m/s^2, until it is travelling at 15 m/s. It then travels at constant speed for 700 m and finally comes to rest in 6 s. Use this data to find the answers to *Problem 5 and 6*.

5 Calculate the tension in the cable supporting the cage during (a) the initial period of acceleration; (b) the period of constant speed travel; (c) the final retardation period.   [(a) 11.6 kN; (b) 19.6 kN; (c) 24.6 kN]

6 A miner having a mass of 80 kg is standing in the cage. Determine the reaction force between the man and the floor of the cage during (a) the initial period of acceleration; (b) the period of constant speed travel; (c) the final retardation period.   [(a) 465 N; (b) 785 N; (c) 985 N]

7 During an experiment, masses of 4 kg and 5 kg are attached to a thread and the thread is passed over a pulley so that both masses hang vertically downwards and are at the same height. When the system is released, find (a) the acceleration of the system; (b) the tension in the thread, assuming no losses in the system.   [(a) 1.09 m/s^2; (b) 43.6 N]

8 Calculate the centripetal force acting on a vehicle of mass 1 t when travelling round a bend of radius 125 m at 40 km/h. If this force should not exceed 750 N, determine the reduction in speed of the vehicle to meet this requirement.   [988 N; 5.1 km/h]

9   A speed-boat negotiates an S-bend consisting of two circular arcs of radii 100 m
    and 150 m. If the speed of the boat is constant at 34 km/h, determine the change
    in acceleration when leaving one arc and entering the other.   [1.49 m/s²]

10 Derive the centripetal force formula $F = \dfrac{mv^2}{r}$, from first principles for an
    object of mass m moving at constant angular velocity, ω, in a circle of radius r on a
    horizontal plane.

11 Friction

A. MAIN POINTS CONCERNING FRICTION

1   When an object, such as a block of wood, is placed on a floor and sufficient
    force is applied to the block, the force being parallel to the floor, the block
    slides across the floor. When the force is removed, motion of the block stops;
    thus there is a force which resists sliding. This force is called **dynamic** or **sliding
    friction**. A force may be applied to the block which is insufficient to move it.
    In this case, the force resisting motion is called the **static friction** or **striction**.
    Thus there are two categories into which a frictional force may be split:
    (i) dynamic or sliding friction force which occurs when motion is taking place, and
    (ii) static friction force which occurs before motion takes place.

2   There are three factors which affect the size and direction of frictional forces.
    (i) The size of the frictional force depends on the type of surface, (a block of
        wood slides more easily on a polished metal surface than on a rough concrete
        surface).
    (ii) The size of the frictional force depends on the size of the force acting at right
        angles to the surfaces in contact, called the **normal force**. Thus, if the weight
        of a block of wood is doubled, the frictional force is doubled when it is
        sliding on the same surface.
    (iii) The direction of the frictional force is always opposite to the direction of
        motion. Thus the frictional force opposes motion, as shown in *Fig 1*.

Fig 1

3   The coefficient of friction, **μ**, is a measure of the amount of friction existing
    between two surfaces. A low value of coefficient of friction indicates that the
    force required for sliding to occur is less than the force required when the coefficient
    of friction is high. The value of the coefficient of friction is given by

$$\mu = \frac{\text{frictional force, } (F)}{\text{normal force, } (N)}$$

Transposing, gives: frictional force = $\mu \times$ normal force,

$$\boxed{F = \mu N}$$

(See *Problems 1 to 3*)

The direction of the forces given in this equation are as shown in *Fig 2*. The coefficient of friction is the ratio of a force to a force, and hence has no units. Typical values for the coefficient of friction when sliding is occurring, i.e., the dynamic coefficient of friction are:

For polished oiled metal surfaces   less than 0.1
For glass on glass   0.4
For rubber on tarmac   close to 1.0

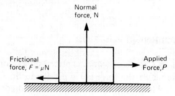

Fig 2

4   In some applications, a low coefficient of friction is desirable, for example, in bearings, pistons moving within cylinders, on ski runs, and so on. However, for such applications as force being transmitted by belt drives and braking systems, a high value of coefficient is necessary. (See *Problems 4 and 5*)

B. WORKED PROBLEMS ON FRICTION

Problem 1 A block of steel requires a force of 10.4 N applied parallel to a steel plate to keep it moving with constant velocity across the plate. If the normal force between the block and the plate is 40 N, determine the dynamic coefficient of friction.

As the block is moving at constant velocity, the force applied must be that required to overcome frictional forces, i.e.,

frictional force, $F = 10.4$ N

The normal force is 40 N, and since $F = \mu N$, (see para. 3),

$$\mu = \frac{F}{N} = \frac{10.4}{40} = 0.26$$

i.e. **the dynamic coefficient of friction is 0.26.**

Problem 2 The surface between the steel block and plate of *Problem 1* is now lubricated and the dynamic coefficient of friction falls to 0.12. Find the new value of force required to push the block at a constant speed.

The normal force depends on the weight of the block and remains unaltered at 40 N. The new value of the dynamic coefficient of friction is 0.12 and since the

128

frictional force $F = \mu N$, $F = 0.12 \times 40 = 4.8$ N. The block is sliding at constant speed, thus the force required to overcome the frictional force is also 4.8 N, i.e.,

the required applied force is 4.8 N

Problem 3 The material of a brake is being tested and it is found that the dynamic coefficient of friction between the material and steel is 0.91. Calculate the normal force when the frictional force is 0.728 kN.

The dynamic coefficient of friction, $\mu = 0.91$

The frictional force, $F = 0.728$ kN $= 728$ N

Since $F = \mu N$, then $N = \dfrac{F}{\mu}$

i.e. normal force, $N = \dfrac{728}{0.91} = 800$ N

i.e. **the normal force is 800 N**

Problem 4 State three advantages and three disadvantages of frictional forces.

Instances where frictional forces are an advantage include:
 (i) Almost all fastening devices rely on frictional forces to keep them in place once secured, examples being screws, nails, nuts, clips and clamps.
 (ii) Satisfactory operation of brakes and clutches rely on frictional forces being present.
(iii) In the absence of frictional forces, most accelerations along a horizontal surface are impossible. For example, a person's shoes just slip when walking is attempted and the tyres of a car just rotate with no forward motion of the car being experienced.

Disadvantages of frictional forces include:
 (i) Energy is wasted in the bearings associated with shafts, axles and gears due to heat being generated.
 (ii) Wear is caused by friction, for example, in shoes, brake lining materials and bearings.
(iii) Energy is wasted when motion through air occurs, (it is much easier to cycle with the wind rather than against it).

Problem 5 Discuss briefly two design implications which arise due to frictional forces and how lubrication may or may not help.

 (i) Bearings are made of an alloy called white metal, which has a relatively low melting point. When the rotating shaft rubs on the white metal bearing, heat is generated by friction, often in one spot and the white metal may melt in this area, rendering the bearing useless. Adequate lubrication, (oil or grease), separates the shaft from the white metal, keeps the coefficient of friction small and prevents damage to the bearing. For very large bearings, oil is pumped under pressure into the bearing and the oil is used to remove the heat generated, often passing through oil coolers before being recirculated. Designers should ensure that the heat generated by friction can be dissipated.
 (ii) Wheels driving belts, to transmit force from one place to another, are used in many workshops. The coefficient of friction between the wheel and the belt

must be high, and it may be increased by dressing the belt with a tar-like substance. Since frictional force is proportional to the normal force, a slipping belt is made more efficient by tightening it, thus increasing the normal and hence the frictional force. Designers should incorporate some belt tension mechanism into the design of such a system.

Problem 6 Explain what is meant by the terms (a) the limiting or static coefficient of friction and (b) the sliding or dynamic coefficient of friction.

(a) When an object is placed on a surface and a force is applied to it in a direction parallel to the surface, if no movement takes place, then the applied force is balanced exactly by the frictional force. As the size of the applied force is increased, a value is reached such that the object is just on the point of moving. The limiting or static coefficient of friction is given by the ratio of this applied force to the normal force, where the normal force is the force acting at right angles to the surfaces in contact.

(b) Once the applied force is sufficient to overcome the striction its value can be reduced slightly and the object moves across the surface. A particular value of the applied force is then sufficient to keep the object moving at a constant velocity. The sliding or dynamic coefficient of friction is the ratio of the applied force, to maintain constant velocity, to the normal force.

C. FURTHER PROBLEMS ON FRICTION

(a) SHORT ANSWER PROBLEMS

1 The . . . . of frictional force depends on the . . . . of surfaces in contact.

2 The . . . . of frictional force depends on the size of the . . . . to the surfaces in contact.

3 The . . . . of frictional force is always . . . . to the direction of motion.

4 The coefficient of friction between surfaces should be a . . . . value for materials concerned with bearings.

5 The coefficient of friction should have a . . . . value for materials concerned with braking systems.

6 The coefficient of dynamic or sliding friction is given by $\dfrac{\ldots\ldots\ldots\ldots\ldots}{\ldots\ldots\ldots\ldots\ldots}$

7 The coefficient of static or limiting friction is given by $\dfrac{\ldots\ldots\ldots\ldots\ldots}{\ldots\ldots\ldots\ldots\ldots}$, when

. . . . is just about to take place.

8 Lubricating surfaces in contact results in a . . . . of the coefficient of friction.

(b) MULTI-CHOICE PROBLEMS (answers on page 173)

Problems 1 to 5 refer to the statements given below. Select the statement required from each group given.

(a) The coefficient of friction depends on the type of surfaces in contact.

130

(b) The coefficient of friction depends on the force acting at right angles to the surfaces in contact.

(c) The coefficient of friction depends on the area of the surfaces in contact.

(d) Frictional force acts in the opposite direction to the direction of motion.

(e) Frictional force acts in the direction of motion.

(f) A low value of coefficient of friction is required between the belt and the wheel in a belt drive system.

(g) A low value of coefficient of friction is required for the materials of a bearing.

(h) The dynamic coefficient of friction is given by $\dfrac{\text{normal force}}{\text{frictional force}}$ at constant speed.

(i) The coefficient of static friction is given by $\dfrac{\text{applied force}}{\text{frictional force}}$ as sliding is just about to start.

(j) Lubrication results in a reduction in the coefficient of friction.

1 Which statement is false from (a), (b), (f) and (i)?

2 Which statement is false from (b), (e), (g) and (j)?

3 Which statement is true from (c), (f), (h) and (i)?

4 Which statement is false from (b), (c), (e) and (j)?

5 Which statement is false from (a), (d), (g) and (h)?

6 The normal force between two surfaces is 100 N and the dynamic coefficient of friction is 0.4. The force required to maintain a constant speed of sliding is:
(a) 100.4 N; (b) 40 N; (c) 99.6 N; (d) 250 N.

7 The normal force between two surfaces is 50 N and the force required to maintain a constant speed of sliding is 25 N. The dynamic coefficient of friction is:
(a) 25; (b) 2; (c) 75; (d) 0.5.

8 The maximum force which can be applied to an object without sliding occurring is 60 N, and the static coefficient of friction is 0.3. The normal force between the two surfaces is:
(a) 200 N; (b) 18 N; (c) 60.3 N; (d) 59.7 N.

(c) CONVENTIONAL PROBLEMS

1 Briefly discuss the factors affecting the size and direction of frictional forces.

2 Name three practical applications where a low value of coefficient of friction is desirable and state briefly how this is achieved in each case.

3 Give three practical applications where a high value of coefficient of friction is required when transmitting forces and discuss how this is achieved.

4 For an object on a surface, two different values of coefficient of friction are possible. Give the names of these two coefficients of friction and state how their values may be obtained.

5 Discuss briefly the effects of frictional force on the design of (a) a hovercraft, (b) a screw and (c) a braking system.

6 The coefficient of friction of a brake pad and a steel disc is 0.82. Determine the normal force between the pad and the disc if the frictional force required is 1025 N.   [1250 N]

7   A force of 0.12 kN is needed to push a bale of cloth along a chute at a constant speed. If the normal force between the bale and the chute is 500 N, determine the dynamic coefficient of friction.   [0.25]

8   The normal force between a belt and its driver wheel is 750 N. If the static coefficient of friction is 0.9 and the dynamic coefficient of friction is 0.87, calculate (a) the maximum force which can be transmitted and (b) maximum force which can be transmitted when the belt is running at a constant speed.

[(a) 675 N; (b) 652.5 N]

12 Work and energy

A. MAIN POINTS CONCERNED WITH WORK AND ENERGY

1 (i) If a body moves as a result of a force being applied to it, the force is said to do
work on the body. The amount of work done is the product of the applied
force and the distance, i.e.
Work done = force × distance moved in the direction of the force
   (ii) The unit of work is the **joule, J,** which is defined as the amount of work done
when a force of one newton acts for a distance of one metre in the direction
of the force. Thus 1 J = 1 Nm.

2 If a graph is plotted of experimental values of force (on the vertical axis) against
distance moved (on the horizontal axis), a force–distance graph or work diagram
is produced. **The area under the graph represents the work done.**

For example, a constant force of 20 N
used to raise a load a height of 8 m
may be represented on a force–
distance graph as shown in *Fig 1(a)*.
The area under the graph shown
shaded, represents the work done.
Hence work done = 20 N × 8 m = **160 J**
Similarly, a spring extended by 20 mm
by a force of 500 N may be represented
by the work diagram shown in *Fig 1(b)*.

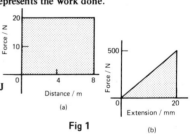

Fig 1

Work done = shaded area = $\frac{1}{2}$ × base × height = $\frac{1}{2}$ × (20 × 10^{-3}) m × 500 N = **5 J**

The work done by a variable force may be found by determining the area enclosed
by the force–distance graph using the trapezoidal rule, the mid-ordinate rule or
Simpson's rule.

3 **Energy** is the capacity, or ability, to do work. The unit of energy is the joule, the
same as for work. Energy is expended when work is done. There are several forms
of energy, and these include mechanical, heat, electrical, chemical, nuclear, light
and sound energy.

4 **Efficiency** is defined as the ratio of the useful output energy to the input energy.
The symbol for efficiency is η (Greek letter eta).

Hence **efficiency** $\eta = \dfrac{\text{useful output energy}}{\text{input energy}}$

Efficiency has no units and is often stated as a percentage. A perfect machine would have an efficiency of 100%. However, all machines have an efficiency lower than this due to friction and other losses. Thus, if the input energy to a motor is 1000 J and the output energy is 800 J then the efficiency is $\frac{800}{1000} \times 100\%$, i.e. 80% (see *Problems 1 to 12*).

5 (i) Mechanical engineering is concerned principally with two kinds of energy, potential energy and kinetic energy.

   (ii) **Potential energy** is energy due to the position of a body.

   The force exerted on a mass of m kg is mg N (where $g = 9.81$ m/s^2, the acceleration due to gravity). When the mass is lifted vertically through a height h m above some datum level, the work done is given by: force $\times$ distance $= (mg)(h)$ J. This work done is stored as potential energy in the mass.

   Hence, **potential energy** $= mg\,h$ **joules** (the potential energy at the datum level being taken as zero).

   (iii) **Kinetic energy** is the energy due to the motion of a body.

   Suppose a force F acts on an object of mass m originally at rest and accelerates it to a velocity v in a distance s.

   Work done = force $\times$ distance $= Fs$

   $\qquad\qquad\qquad\qquad\qquad = (ma)(s)$, (if no energy is lost), where a is the acceleration.

   However $v^2 = 2as$, from which $a = \dfrac{v^2}{2s}$

   Hence work done $= m\left(\dfrac{v^2}{2s}\right)s = \dfrac{1}{2}m\,v^2$

   This energy is called the kinetic energy of the mass m.

   i.e. **kinetic energy** $= \frac{1}{2}mv^2$ **joules**

6 (i) Energy may be converted from one form to another. **The principle of conservation of energy** states that the total amount of energy remains the same in such conversions, i.e. energy cannot be created or destroyed.

   (ii) In mechanics, the potential energy possessed by a body is frequently converted into kinetic energy, and vice versa. When a mass is falling freely, its potential energy decreases as it loses height, and its kinetic energy increases as its velocity increases. Ignoring air frictional losses, at all times:

   Potential energy + kinetic energy = a constant

   (iii) If friction is present, then work is done overcoming the resistance due to friction and this is dissipated as heat. Then,

   Initial energy = final energy + work done overcoming frictional resistance

   (iv) Kinetic energy is not always conserved in collisions. Collisions in which kinetic energy is conserved (i.e. stays the same) are called **elastic collisions**, and those in which it is not conserved are termed **inelastic collisions**.
   (See *Problems 13 to 17*)

7 (i) **Heat** is a form of energy and is measured in joules.

   (ii) **Temperature** is the degree of hotness or coldness of a substance. Heat and temperature are thus **not** the same thing. For example, twice the heat energy is needed to boil a full container of water than half a container − that is, different amounts of heat energy are needed to cause an equal rise in the temperature of different amounts of the same substance.

8   Temperature is measured either (i) on the **Celsius (°C) scale** (formerly Centigrade), where the temperature at which ice melts, i.e. the freezing point of water, is taken as 0°C and the point at which water boils under normal atmospheric pressure is taken as 100°C, or (ii) on the **thermodynamic scale**, in which the unit of temperature is the kelvin (K). The kelvin scale uses the same temperature interval as the Celsius scale but as its zero takes the 'absolute zero of temperature' which is at about−273°C. Hence kelvin temperature = degree Celsius + 273

i.e. $K = (°C) + 273$

Thus, for example, $0°C = 273$ K, $25°C = 298$ K and $100°C = 373$ K.
(See *Problems 18 and 19*.)

9   A **thermometer** is an instrument which measures temperature. Any substance which possesses one or more properties which vary with temperature can be used to measure temperature. These properties include changes in length, area or volume, electrical resistance or in colour. Examples of temperature-measuring devices include liquid-in-glass thermometers, thermocouples, resistance thermometers and pyrometers.

10 (i) The **specific heat capacity** of a substance is the quantity of heat energy required to raise the temperature of 1 kg of the substance by 1°C.
(ii) The symbol used for specific heat capacity is c and the units are J/(kg °C) or J/(kg K). Note that these units may also be written as J kg^{-1} °C^{-1} or J kg^{-1} K^{-1}.
(iii) Some typical values of specific heat capacity for the range of temperature 0°C to 100°C include:

Water 4190 J/(kg °C)   Ice 2100 J/(kg °C)
Aluminium, 950 J/(kg °C)   Copper 390 J/(kg °C)
Iron, 500 J/(kg °C)   Lead 130 J/(kg °C)

Hence to raise the temperature of 1 kg of iron by 1°C requires 500 J of energy; to raise the temperature of 5 kg of iron by 1°C requires (500 × 5) J of energy; and to raise the temperature of 5 kg of iron by 40°C requires (500 × 5 × 40) J of energy, i.e. 100 kJ.

In general, the quantity of heat energy Q, required to raise a mass m kg of a substance with a specific heat capacity c J/(kg°C) from temperature t_1 °C to t_2 °C is given by:

$$Q = mc(t_2 - t_1) \text{ joules}$$

(See *Problems 20 to 25*)

11 A material may exist in any one of three states − solid, liquid or gas. If heat is supplied at a constant rate to some ice, initially at, say −30°C, its temperature rises as shown in *Figure 2*. Initially the temperature increases from −30°C to 0°C as shown by the line AB. It then remains constant at 0°C for the time BC required for the ice to melt into water. When melting commences the energy gained by continual heating is offset by the energy required for the change of state and the temperature remains constant even though heating is continued. When the ice is completely melted to water, continual heating raises the temperature to 100°C, as shown by CD in *Fig 2*. The water then begins to boil and the temperature again remains constant at 100°C, shown as DE, until all the water has vaporised. Continual

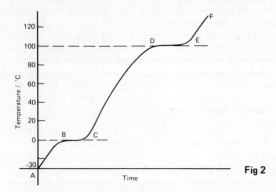

Fig 2

heating raises the temperature of the steam as shown by EF in the region where the steam is termed superheated.

Changes of state from solid to liquid or liquid to gas occur without change of temperature and such changes are reversible processes. When heat energy flows to or from a substance and causes a change of temperature, such as between A and B, between C and D and between E and F in *Fig 2*, it is called **sensible heat** (since it can be 'sensed' by a thermometer).

Heat energy which flows to or from a substance while the temperature remains constant, such as between B and C and between D and E in *Fig 2*, is called **latent heat** (latent means concealed or hidden).

12 (i) The **specific latent heat of fusion** is the heat required to change 1 kg of a substance from the solid state to the liquid state (or vice versa) at constant temperature.

(ii) The **specific latent heat of vaporisation** is the heat required to change 1 kg of a substance from a liquid to a gaseous state (or vice versa) at constant temperature.

(iii) The units of the specific latent heats of fusion and vaporisation are J/kg, or more often, kJ/kg, and some typical values are shown below.

	Latent heat of fusion (kJ/kg)	Melting point (°C)
Mercury	11.8	−39
Lead	22	327
Silver	100	957
Ice	335	0
Aluminium	387	660

	Latent heat of vaporisation (kJ/kg)	Boiling point (°C)
Oxygen	214	−183
Mercury	286	357
Ethyl alcohol	857	79
Water	2257	100

(iv) The quantity of heat Q supplied or given out during a change of state is given by:

$$Q = mL$$

where m is the mass in kilograms and L is the specific latent heat. Thus, for example, the heat required to convert 10 kg of ice at 0°C to water at 0°C is given by:

10 kg × 335 kJ/kg, i.e. 3350 kJ or 3.35 MJ.

(See *Problems 26 to 31*)

B. WORKED PROBLEMS ON WORK AND ENERGY

Problem 1 Calculate the work done when a force of 40 N pushes an object a distance of 500 m in the same direction as the force.

Work done = force × distance moved in the direction of the force
           = 40 N × 500 m = 20 000 J (since 1 J = 1 Nm)
i.e. work done = **20 kJ**

Problem 2 Calculate the work done when a mass is lifted vertically by a crane to a height of 5 m, the force required to lift the mass being 98 N.

When work is done in lifting then: Work done = (weight of the body) ×·(vertical distance moved).
Weight is the downward force due to the mass of an object.
Hence work done = 98 N × 5 m = **490 J**

Problem 3 A motor supplies a constant force of 1 kN which is used to move a load a distance of 5 m. The force is then changed to a constant 500 N and the load is moved a further 15 m. Draw the force–distance graph for the operation and from the graph determine the total work done by the motor.

The force–distance graph or work diagram is shown in *Fig 3.*
Between points A and B a constant force of 1000 N moves the load 5 m; between points C and D a constant force of 500 N moves the load from 5 m to 20 m.
Total work done = area under the force–distance graph = area ABFE + area CDGF
= (1000 N × 5 m) + (500 N × 15 m)
= 5000 J + 7500 J = 12 500 J = **12.5 kJ**

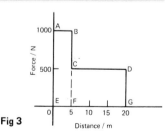

Fig 3

Problem 4 A spring initially in a relaxed state is extended by 100 mm. Determine the work done by using a work diagram if the spring requires a force of 0.6 N per mm of stretch.

Force required for a 100 mm extension
= 100 mm × 0.6 N mm^{-1} = 60 N. *Fig 4*
shows the force–extension graph or work
diagram representing the increase in extension
in proportion to the force, as the force is
increased from 0 to 60 N. The work done is
the area under the graph (shown shaded).
Hence work done = $\frac{1}{2}$ × base × height
$= \frac{1}{2}$ × 100 mm × 60 N
$= \frac{1}{2}$ × 100 × 10^{-3} m × 60 N = **3 J**

Fig 4

(Alternatively, average force during extension = $\dfrac{60 - 0}{2}$ = 30 N and
total extension = 100 mm = 0.1 m.
Hence work done = average force × extension = 30 N × 0.1 m = **3 J**)

Problem 5 A spring requires a force of 10 N to cause an extension of 50 mm.
Determine the work done in extending the spring (a) from zero to 30 mm, and
(b) from 30 mm to 50 mm.

Figure 5 shows the force–extension graph for the spring.
(a) Work done in extending the spring from zero to 30 mm is given by area ABO
of *Fig 5*, i.e.

Work done = $\frac{1}{2}$ × base × height = $\frac{1}{2}$ × 30 × 10^{-3} m × 6 N = 90 × 10^{-3} J = **0.09 J**

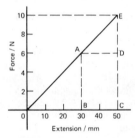

Fig 5

(b) Work done in extending the spring from 30 mm to 50 mm is given by:

Area ABCE of *Fig 5*, i.e. work done = area ABCD + area ADE
= (20 × 10^{-3} m × 6 N) + $\frac{1}{2}$ (20 × 10^{-3} m)(4N) = 0.12 J + 0.04 J = **0.16 J**

Problem 6 Calculate the work done when a mass of 20 kg is lifted vertically
through a distance of 5.0 m.

The force to be overcome when lifting a mass of 20 kg vertically upwards is *mg*,
i.e. 20 × 9.81 = 196.2 N.
Work done = force × distance = 196.2 × 5.0 = **981 J**

Problem 7 Water is pumped vertically upwards through a distance of 50.0 m and
the work done is 294.3 kJ. Determine the number of litres of water pumped.
(1 litre of water has a mass of 1 kg.)

Work done = force × distance, i.e. 294 300 = force × 50.0,

from which force = $\dfrac{294\ 300}{50.0}$ = 5886 N.

The force to be overcome when lifting a mass m kg vertically upwards is mg, i.e. $(m \times 9.81)$ N.

Thus 5886 = $m \times$ 9.81, from which mass, $m = \dfrac{58.86}{9.81}$ = 600 kg.

Since 1 litre of water has a mass of 1 kg, **600 litres of water are pumped**

Problem 8 The force on the cutting tool of a shaping machine varies over the length of cut as follows:

Distance (mm)	0	20	40	60	80	100
Force (kN)	60	72	65	53	44	50

Determine the work done as the tool moves through a distance of 100 mm.

The force–distance graph for the given data is shown in *Fig 6*. The work done is given by the area under the graph. this area may be determined by an approximate method.

Using the mid-ordinate rule, with each strip of width 20 mm, mid-ordinates are erected as shown, y_1, y_2, y_3, y_4 and y_5 and each is measured.

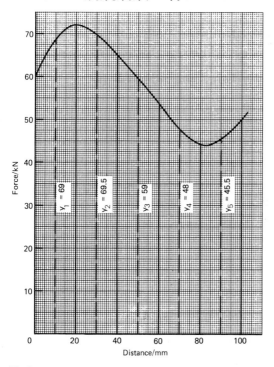

Fig 6

Area under curve = (width of each strip)(sum of mid-ordinate values)

= (20)(69 + 69.5 + 59 + 48 + 45.5) = (20)(291)

= 5820 kN mm = 5820 Nm = 5820 J.

Hence the work done as the tool moves through 100 mm is **5.82 kJ**

Problem 9 A machine exerts a force of 200 N in lifting a mass through a height of 6 m. If 2 kJ of energy are supplied to it, what is the efficiency of the machine?

Work done in lifting mass = force × distance moved

= weight of body × distance moved

= 200 N × 6 m = 1200 J = useful energy output

Energy input = 2 kJ = 2000 J.

Efficiency $\eta = \dfrac{\text{useful output energy}}{\text{input energy}} = \dfrac{1200}{2000} = $ **0.6 or 60%**

Problem 10 Calculate the useful output energy of an electric motor which is 70% efficient if it uses 600 J of electrical energy.

Efficiency $\eta = \dfrac{\text{useful output energy}}{\text{input energy}}$, thus $\dfrac{70}{100} = \dfrac{\text{output energy}}{600 \text{ J}}$

from which, output energy = $\dfrac{70}{100} \times 600 = $ **420 J**

Problem 11 4 kJ of energy are supplied to a machine used for lifting a mass. The force required is 800 N. If the machine has an efficiency of 50%, to what height will it lift the mass?

Efficiency $\eta = \dfrac{\text{output energy}}{\text{input energy}}$, i.e. $\dfrac{50}{100} = \dfrac{\text{output energy}}{4000 \text{ J}}$

from which, output energy = $\dfrac{50}{100} \times 4000 = 2000$ J.

Work done = force × distance moved, hence 2000 J = 800 N × height

from which, height = $\dfrac{2000 \text{ J}}{800 \text{ N}} = $ **2.5 m**

Problem 12 A hoist exerts a force of 500 N in raising a load through a height of 20 m. The efficiency of the hoist gears is 75% and the efficiency of the motor is 80%. Calculate the input energy to the hoist.

The hoist system is shown diagrammatically in *Fig 7*.

Output energy = work done = force × distance = 500 N × 20 m = 10 000 J

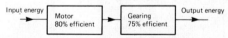

Fig 7

For the gearing, efficiency = $\dfrac{\text{output energy}}{\text{input energy}}$, i.e. $\dfrac{75}{100} = \dfrac{10\,000}{\text{input energy}}$

from which, the input energy to the gears = $10\,000 \times \dfrac{100}{75} = $ **13 333 J**

The input energy to the gears is the same as the output energy of the motor.

140

Thus, for the motor, efficiency = $\dfrac{\text{output energy}}{\text{input energy}}$, i.e. $\dfrac{80}{100} = \dfrac{13\,333}{\text{input energy}}$

Hence the input energy to the system = $13\,333 \times \dfrac{100}{80} = 16\,670$ J = **16.67 kJ**

Problem 13 A car of mass 800 kg is climbing an incline at 10° to the horizontal. Determine the increase in potential energy of the car as it moves a distance of 50 m up the incline.

With reference to *Fig 8*, sin 10° = $\dfrac{h}{50}$, from which, $h = 50 \sin 10° = 8.682$ m.

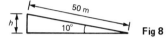

Fig 8

Hence increase in potential energy = $mgh = 800$ kg $\times$ 9.81 m/s^2 $\times$ 8.682 m
$= 68\,140$ J or 68.14 kJ

Problem 14 At the instant of striking, a hammer of mass 30 kg has a velocity of 15 m/s. Determine the kinetic energy in the hammer.

Kinetic energy = $\frac{1}{2} mv^2 = \frac{1}{2}$ (30 kg)(15 m/s)^2
i.e. kinetic energy in hammer = **3375 J**.

Problem 15 A lorry having a mass of 1.5 t is travelling along a level road at 72 km/h. When the brakes are applied, the speed decreases to 18 km/h. Determine how much the kinetic energy of the lorry is reduced.

Initial velocity of lorry $v_1 = 72$ km/h $= 72 \dfrac{\text{km}}{\text{h}} \times 1000 \dfrac{\text{m}}{\text{km}} \times \dfrac{1\,\text{h}}{3600\,\text{s}} = \dfrac{72}{3.6} = 20$ m/s

Final velocity of lorry $v_2 = \dfrac{18}{3.6} = 5$ m/s

Mass of lorry, $m = 1.5$ t $= 1500$ kg

Initial kinetic energy of the lorry = $\frac{1}{2} mv_1{}^2 = \frac{1}{2}$ (1500)(20)$^2 = 300$ kJ
Final kinetic energy of the lorry = $\frac{1}{2} mv_2{}^2 = \frac{1}{2}$ (1500)(5)$^2 = 18.75$ kJ
Hence the change in kinetic energy = $300 - 18.75 =$ **281.25 kJ**
(Part of this reduction in kinetic energy is converted into heat energy in the brakes of the lorry and is hence dissipated in overcoming frictional forces and air friction.)

Problem 16 A canister containing a meteorology balloon of mass 4 kg is fired vertically upwards from a gun with an initial velocity of 400 m/s. Neglecting the air resistance, calculate (a) its initial kinetic energy; (b) its velocity at a height of 1 km; (c) the maximum height reached.

(a) Initial kinetic energy = $\frac{1}{2} mv^2 = \frac{1}{2}$ (4)(400)^2 = **320 kJ**

(b) At a height of 1 km, potential energy = $mgh = 4 \times 9.81 \times 1000 = 39.24$ kJ.
By the principle of conservation of energy:
Potential energy + kinetic energy at 1 km = initial kinetic energy

Hence $39\ 240 + \frac{1}{2}\ mv^2 = 320\ 000$

from which, $\frac{1}{2}\ (4)\ v^2 = 320\ 000 - 39\ 240 = 280\ 760$

Hence, $v = \sqrt{\left(\frac{2 \times 280\ 760}{4}\right)} = 374.7$ m/s,

i.e. **the velocity of the canister at a height of 1 km is 374.7 m/s**

(c) At the maximum height, the velocity of the canister is zero and all the kinetic energy has been converted into potential energy.

Hence potential energy = initial kinetic energy = 320 000 J from part (a).
Then $320\ 000 = mgh = (4)(9.81)\ h$

from which, height $h = \dfrac{320\ 000}{(4)(9.81)} = 8155$ m,

i.e. **the maximum height reached is 8155 m**

Problem 17 A pile-driver of mass 500 kg falls freely through a height of 1.5 m on to a pile of mass 200 kg. Determine the velocity with which the driver hits the pile. If, at impact, 3 kJ of energy are lost due to heat and sound, the remaining energy being possessed by the pile and driver as they are driven together into the ground a distance of 200 mm, determine (a) the common velocity immediately after impact; (b) the average resistance of the ground.

The potential energy of the pile-driver is converted into kinetic energy,

Thus potential energy = kinetic energy, i.e. $mgh = \frac{1}{2}\ mv^2$
from which, velocity $v = \sqrt{(2gh)} = \sqrt{[(2)(9.81)(1.5)]} = 5.42$ m/s
Hence the pile-driver hits the pile at a velocity of **5.42 m/s**

(a) Before impact, kinetic energy of pile driver $= \frac{1}{2}\ mv^2 = \frac{1}{2}\ (500)(5.42)^2 = 7.34$ kJ
Kinetic energy after impact $= 7.34 - 3 = 4.34$ kJ.
Thus the pile-driver and pile together have a mass of $500 + 200 = 700$ kg and possess kinetic energy of 4.34 kJ.

Hence $4.34 \times 10^3 = \frac{1}{2}\ mv^2 = \frac{1}{2}\ (700)\ v^2$

from which, velocity $v = \sqrt{\left(\frac{2 \times 4.34 \times 10^3}{700}\right)} = 3.52$ m/s.

Thus the common velocity after impact is **3.52 m/s**

(b) The kinetic energy after impact is absorbed in overcoming the resistance of the ground, in a distance of 200 mm.

Kinetic energy = work done = resistance × distance
i.e. $4.34 \times 10^3 =$ resistance × 0.200

from which, resistance $= \dfrac{4.34 \times 10^3}{0.200} = 21\ 700$ N

Hence the average resistance of the ground is **21.7 kN**

Problem 18 Convert the following temperatures into the kelvin scale:
(a) 37°C; (b) −28°C.

From para. 8, kelvin temperature = degree Celsius + 273
(a) 37°C corresponds to a kelvin temperature of $37 + 273$, i.e. **310 K.**
(b) −28°C corresponds to a kelvin temperature of $-28 + 273$, i.e. **245 K.**

Problem 19 Convert the following temperatures into the Celsius scale:
(a) 365 K; (b) 213 K.

From para. 8, K = (°C) + 273. Hence, degree Celsius = kelvin temperature − 273.
(a) 365 K corresponds to 365 − 273, i.e. **92°C**.
(b) 213 K corresponds to 213 − 273, i.e. **−60°C**.

Problem 20 Calculate the quantity of heat required to raise the temperature of 5 kg of water from 0°C to 100°C. Assume the specific heat capacity of water is 4200 J/(kg °C).

Quantity of heat energy, $Q = mc(t_2 - t_1)$, from para. 10,
$$= 5 \text{ kg} \times 4200 \text{ J/(kg °C)} \times (100 - 0)°C$$
$$= 5 \times 4200 \times 100 = 2\,100\,000 \text{ J or } 2100 \text{ kJ or } 2.1 \text{ MJ}$$

Problem 21 A block of cast iron having a mass of 10 kg cools from a temperature of 150°C to 50°C. How much energy is lost by the cast iron? Assume the specific heat capacity of copper is 500 J/(kg °C).

Quantity of heat energy, $Q = mc(t_2 - t_1) = 10 \text{ kg} \times 500 \text{ J/(kg °C)} \times (50 - 150)°C$
$$= 10 \times 500 \times (-100)$$
$$= -500\,000 \text{ J or } -500 \text{ kJ or } -0.5 \text{ MJ}$$
(Note that the minus sign indicates that heat is given out or lost.)

Problem 22 Some lead having a specific heat capacity of 130 J/(kg °C) is heated from 27°C to its melting point at 327°C. If the quantity of heat required is 780 kJ, determine the mass of the lead.

Quantity of heat, $Q = mc(t_2 - t_1)$
Hence 780×10^3 J $= m \times 130$ J/(kg °C) $\times (327 - 27)°C$
i.e. 780 000 $= m \times 130 \times 300$

from which, mass $m \doteq \dfrac{780\,000}{130 \times 300}$ kg $= 20$ kg

Problem 23 273 kJ of heat energy are required to raise the temperature of 10 kg of copper from 15°C to 85°C. Determine the specific heat capacity of copper.

Quantity of heat, $Q = mc(t_2 - t_1)$
Hence, 273×10^3 J $= 10$ kg $\times c \times (85 - 15)°C$, where $c =$ specific heat capacity
i.e.   273 000 $= 10 \times c \times 70$

from which, specific heat capacity of copper, $c = \dfrac{273\,000}{10 \times 70} = 390$ J/(kg °C)

Problem 24 5.7 MJ of heat energy are supplied to 30 kg of aluminium which is initially at a temperature of 20°C. If the specific heat capacity of aluminium is 950 J/(kg °C), determine its final temperature.

Quantity of heat, $Q = mc(t_2 - t_1)$
Hence 5.7 $\times 10^6$ J $= 30$ kg $\times 950$ J/(kg °C) $\times (t_2 - 20)°C$

from which, $(t_2 - 20) = \dfrac{5.7 \times 10^6}{30 \times 950} = 200°C$

Hence the final temperature, $t_2 = 200 + 20 = \mathbf{220°C}$

Problem 25 A copper container of mass 500 g contains 1 litre of water at 293 K. Calculate the quantity of heat required to raise the temperature of the water and container to boiling point assuming there are no heat losses. Assume that the specific heat capacity of copper is 390 J/(kg K), the specific heat capacity of water is 4.2 kJ/(kg K) and 1 litre of water has a mass of 1 kg.

Heat is required to raise the temperature of the water, and also to raise the temperature of the copper container.

For the water: $m = 1$ kg; $t_1 = 293$ K; $t_2 = 373$ K (i.e. boiling point); $c = 4.2$ kJ/(kg K)

Quantity of heat required for the water, $Q_W = mc\,(t_2 - t_1)$
$$= (1\ \text{kg})(4.2\ \tfrac{\text{kJ}}{\text{kg K}})(373 - 293)\ \text{K}$$
$$= 4.2 \times 80\ \text{kJ}$$

i.e. $Q_W = 336$ kJ

For the copper container: $m = 500$ g $= 0.5$ kg; $t_1 = 293$ K; $t_2 = 373$ K; $c = 390$ J/(kg K) $= 0.39$ kJ/(kg K).

Quantity of heat required for the copper container, $Q_C = mc(t_2 - t_1)$,
$$= (0.5\ \text{kg})(0.39\ \text{kJ/(kg K)})(80\ \text{K})$$

i.e. $Q_C = 15.6$ kJ

Total quantity of heat required, $Q = Q_W + Q_C = 336 + 15.6 = \mathbf{351.6\ kJ}$

Problem 26 Steam initially at a temperature of 130°C is cooled to a temperature of 20°C below the freezing point of water, the loss of heat energy being at a constant rate. Make a sketch, and briefly explain, the expected temperature–time graph representing this change.

A temperature–time graph representing the change is shown in *Fig 9*. Initially steam cools until it reaches the boiling point of water at 100°C. Temperature then remains constant, i.e. between A and B, even though it is still giving off heat (i.e. latent heat). When all the steam at 100°C has changed to water at 100°C it starts to cool again until it reaches the freezing point of water at 0°C. From C to D the

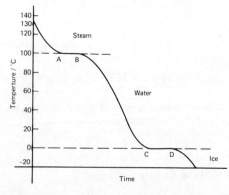

Fig 9

temperature again remains constant until all the water is converted to ice. The temperature of the ice then decreases as shown.

Problem 27 How much heat is needed to completely melt 12 kg of ice at 0°C? Assume the latent heat of fusion of ice is 335 kJ/kg.

Quantity of heat required, $Q = mL$, from para. 11,
$$= 12 \text{ kg} \times 335 \text{ kJ/kg} = \textbf{4020 kJ} \text{ or } \textbf{4.02 MJ}.$$

Problem 28 Calculate the heat required to convert 5 kg of water at 100° C to steam at 100°C. Assume the latent heat of vaporisation of water is 2260 kJ/kg.

Quantity of heat required, $Q = mL$
$$= 5 \text{ kg} \times 2260 \text{ kJ/kg} = \textbf{11 300 kJ} \text{ or } \textbf{11.3 MJ}$$

Problem 29 Determine the heat energy needed to convert 5 kg of ice initially at −20°C completely to water at 0°C. Assume the specific heat capacity of ice is 2100 J/(kg °C) and the specific latent heat of fusion of ice is 335 kJ/kg.

Quantity of heat energy needed, Q = sensible heat + latent heat.
The quantity of heat needed to raise the temperature of ice from −20°C to 0°C,
i.e. sensible heat, $Q_1 = mc(t_2 - t_1) = 5 \text{ kg} \times 2100 \text{ J/(kg °C)} \times (0 - -20)°\text{C}$
$$= (5 \times 2100 \times 20) \text{ J} = 210 \text{ kJ}.$$
The quantity of heat needed to melt 5 kg of ice at 0°C,
i.e. the latent heat, $Q_2 = mL = 5 \text{ kg} \times 335 \text{ kJ/kg} = 1675 \text{ kJ}$
Total heat energy needed, $Q = Q_1 + Q_2 = 210 + 1675 = \textbf{1885 kJ}$

Problem 30 Calculate the heat energy required to convert completely 10 kg of water at 50°C into steam at 100°C, given that the specific heat capacity of water is 4200 J/(kg °C) and the specific latent heat of vaporisation of water is 2260 kJ/kg.

Quantity of heat required = sensible heat + latent heat
Sensible heat, $Q_1 = mc (t_2 - t_1) = 10 \text{ kg} \times 4200 \text{ J/(kg °C)} \times (100-50)°\text{C} = 2100 \text{ k}$
Latent heat, $Q_2 = mL = 10 \text{ kg} \times 2260 \text{ kJ/kg} = 22 600 \text{ kJ}$
Total heat energy required, $Q = Q_1 + Q_2 = (2100 + 22 600) \text{ kJ}$
$$= \textbf{24 700 kJ} \text{ or } \textbf{24.70 MJ}$$

Problem 31 Determine the amount of heat energy needed to change 400 g of ice, initially at −20°C, into steam at 120°C. Assume the following: latent heat of fusion of ice = 335 kJ/kg; latent heat of vaporisation of water = 2260 kJ/kg; specific heat capacity of ice = 2.14 kJ/(kg °C); specific heat capacity of water = 4.2 kJ/(kg °C); specific heat capacity of steam = 2.01 kJ/(kg °C).

The energy needed is determined in five stages:
(i) Heat energy needed to change the temperature of ice from −20°C to 0°C is given by:

$$Q_1 = mc(t_2 - t_1) = 0.4 \text{ kg} \times 2.14 \text{ kJ/(kg °C)} \times (0 - -20)°\text{C} = 17.12 \text{ kJ}.$$

(ii) Latent heat needed to change ice at 0°C into water at 0°C is given by:

$$Q_2 = mL_f = 0.4 \text{ kg} \times 335 \text{ kJ/kg} = 134 \text{ kJ}$$

145

(iii) Heat energy needed to change the temperature of water from $0°C$ (i.e. melting point) to $100°C$ (i.e. boiling point) is given by:

$$Q_3 = mc(t_2 - t_1) = 0.4 \text{ kg} \times 4.2 \text{ kJ/(kg °C)} \times 100°C = 168 \text{ kJ}.$$

(iv) Latent heat needed to change water at $100°C$ into steam at $100°C$ is given by:

$$Q_4 = mL_V = 0.4 \text{ kg} \times 2260 \text{ kJ/kg} = 904 \text{ kJ}.$$

(v) Heat energy needed to change steam at $100°C$ into steam at $120°C$ is given by:

$$Q_5 = mc(t_2 - t_1) = 0.4 \text{ kg} \times 2.01 \text{ kJ/(kg °C)} \times 20°C = 16.08 \text{ kJ}.$$

Total heat energy needed, $Q = Q_1 + Q_2 + Q_3 + Q_4 + Q_5$
$$= 17.12 + 134 + 168 + 904 + 16.08 = \mathbf{1239.2 \text{ kJ}}$$

C. FURTHER PROBLEMS ON WORK AND ENERGY

(a) SHORT ANSWER PROBLEMS

1 Define work in terms of force applied and distance moved.

2 Define energy, and state its unit.

3 Define the joule.

4 The area under a force–distance graph represents . . . .

5 Name five forms of energy.

6 (a) Define efficiency in terms of energy input and energy output.
(b) State the symbol used for efficiency.

7 Define potential energy.

8 The change in potential energy of a body of mass m kg when lifted vertically upwards to a height h m is given by . . . .

9 What is kinetic energy?

10 The kinetic energy of a body of mass m kg and moving at a velocity of v m/s is given by . . . .

11 State the principle of conservation of energy.

12 Distinguish between elastic and inelastic collisions.

13 Differentiate between temperature and heat.

14 Name two scales on which temperature is measured.

15 How are the fixed points on the Celsius scale obtained?

16 Define specific heat capacity and name its unit.

17 Differentiate between sensible and latent heat.

18 The quantity of heat, Q, required to raise a mass m kg from temperature $t_1 °C$ to $t_2 °C$, the specific heat capacity being c, is given by $Q = $ . . . .

19 What is meant by the specific latent heat of fusion?

20 Define the specific latent heat of vaporisation.

1 An object is lifted 2000 mm by a crane. If the force required is 100 N, the work done is:

(a) $\frac{1}{20}$ N; (b) 200 kN; (c) 200 N; (d) 20 N.

2 A motor having an efficiency of 0.8 uses 800 J of electrical energy. The output energy of the motor is:
(a) 800 J; (b) 1000 J; (c) 640 J.

3 6 kJ of work is done by a force in moving an object uniformly through 120 m. The force applied is:
(a) 50 N; (b) 20 N; (c) 720 N; (d) 12 N.

4 A force–extension graph for a spring is shown in *Fig 10*. Which of the following statements is false? The work done in extending the spring:
(a) from 0 to 100 mm is 5 N;
(b) from 0 to 50 mm is 1.25 N;
(c) from 20 mm to 60 mm is 1.6 N;
(d) from 60 mm to 100 mm is 3.75 N.

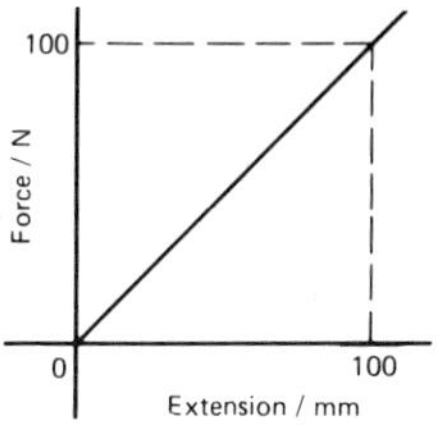

Fig 10

5 A vehicle of mass 1 tonne climbs an incline of 30° to the horizontal. Taking the acceleration due to gravity as 10 m/s², the increase in potential energy of the vehicle as it moves a distance of 200 m up the incline is:
(a) 1 kJ; (b) 2 MJ; (c) 1 MJ; (d) 2 kJ.

6 A bullet of mass 100 g is fired from a gun with an initial velocity of 360 km/h. Neglecting air resistance, the initial kinetic energy possessed by the bullet is:
(a) 6.48 kJ; (b) 500 J; (c) 500 kJ. (d) 6.48 MJ.

7 Heat energy is measured in:
(a) kelvin; (b) watts; (c) kilograms; (d) joules.

8 A change of temperature of 20°C is equivalent to a change of thermodynamic temperature of: (a) 293 K; (b) 20 K.

9 A temperature of 20°C is equivalent to: (a) 293 K; (b) 20 K.

10 The unit of specific heat capacity is: (a) joules per kilogram; (b) joules; (c) joules per kilogram kelvin; (c) cubic metres.

11 The quantity of heat required to raise the temperature of 500 g of iron by 2°C, given that the specific heat capacity is 500 J/(kg °C), is:
(a) 50 kJ; (b) 0.5 kJ; (c) 2 J; (d) 250 kJ.

12 The heat energy required to change 1 kg of a substance from a liquid to a gaseous state at the same temperature is called:
(a) specific heat capacity; (b) specific latent heat of vaporisation; (c) sensible heat; (d) specific latent heat of fusion.

(c) CONVENTIONAL PROBLEMS

(where necessary, take *g* as 9.81 m/s²)

1 Determine the work done when a force of 50 N pushes an object 1.5 km in the same direction as the force. [75 kJ]

2   Calculate the work done when a mass of weight 200 N is lifted vertically by a crane to a height of 100 m. [20 kJ]

3   A motor supplies a constant force of 2 kN to move a load 10 m. The force is then changed to a constant 1.5 kN and the load is moved a further 20 m. Draw the force–distance graph for the complete operation, and, from the graph, determine the total work done by the motor. [50 kJ]

4   A spring, initially relaxed, is extended 80 mm. Draw a work diagram and hence determine the work done if the spring requires a force of 0.5 N/mm of stretch. [1.6 J]

5   A spring requires a force of 50 N to cause an extension of 100 mm. Determine the work done in extending the spring (a) from 0 to 100 mm; (b) from 40 mm to 100 mm [(a) 2.5 J; (b) 2.1 J]

6   The resistance to a cutting tool varies during the cutting stroke of 800 mm as follows:
    (i) The resistance increases uniformly from an initial 5000 N to 10 000 N as the tool moves 500 mm.
    (ii) The resistance falls uniformly from 10 000 N to 6000 N as the tool moves 300 mm. Draw the work diagram and calculate the work done in one cutting stroke. [6.15 kJ]

7   Determine the work done when a mass of 25 kg is lifted vertically through a height of 8.0 m. [1.962 kJ]

8   Water is pumped vertically upwards through a distance of 30 m and the work done is 132.4 kJ. Determine the number of litres of water pumped (1 litre of water has a mass of 1 kg). [450 litres]

9   A lorry is moving away from rest and the force exerted by the engine varies with distance as follows:

Distance (m)	0	10	20	30	40	50
Force (N)	300	280	260	233	190	150

Determine the work done as the lorry moves from rest through a distance of 50 m. [12 kJ]

10 A force is applied to a mass. The variation of force with distance moved is as follows:

Distance (mm)	0	1	2	3	4	5	6	7	8
Force (N)	0	14	27	45	47	43	37	19	0

Draw the force–distance diagram and hence determine the work done by the force when moving the mass through a distance of 8 mm. [0.235 J]

11 A machine lifts a mass of weight 490.5 N through a height of 12 m when 7.85 kJ of energy is supplied to it. Determine the efficiency of the machine. [75%]

12 Determine the output energy of an electric motor which is 60% efficient if it uses 2 kJ of electrical energy. [1.2 kJ]

13 A machine which is used for lifting a particular mass is supplied with 5 kJ of energy. If the machine has an efficiency of 65% and exerts a force of 812.5 N, to what height will it lift the mass? [4 m]

14 A load is hoisted 42 m and requires a force of 100 N. The efficiency of the hoist gear is 60% and that of the motor is 70%. Determine the input energy to the hoist. [10 kJ]

15 An object of mass 400 g is thrown vertically upwards and its maximum increase in potential energy is 32.6 J. Determine the maximum height reached, neglecting air resistance. [8.31 m]

16 A ball bearing of mass 100 g rolls down from the top of a chute of length 400 m inclined at an angle of 30° to the horizontal. Determine the decrease in potential energy of the ball bearing as it reaches the bottom of the chute. [196.2 J]

17 A vehicle of mass 800 kg is travelling at 54 km/h when its brakes are applied. Find the kinetic energy lost when the car comes to rest. [90 kJ]

18 Supplies of mass 300 kg are dropped from a helicopter flying at an altitude of 60 m. Determine the potential energy of the supplies relative to the ground at the instant of release, and its kinetic energy as it strikes the ground. [176.6 kJ; 176.6 kJ]

19 A shell of mass 10 kg is fired vertically upwards with an initial velocity of 200 m/s. Determine its initial kinetic energy and the maximum height reached, correct to the nearest metre, neglecting air resistance. [200 kJ; 2039 m]

20 The potential energy of a mass is increased by 20.0 kJ when it is lifted vertically through a height of 25.0 m. It is now released and allowed to fall freely. Neglecting air resistance, find its kinetic energy and its velocity after it has fallen 10.0 m. [8 kJ; 14.0 m/s]

21 A pile driver of mass 400 kg falls freely through a height of 1.2 m onto a pile of mass 150 kg. Determine the velocity with which the driver hits the pile. If, at impact, 2.5 kJ of energy are lost due to heat and sound, the remaining energy being possessed by the pile and driver as they are driven together into the ground a distance of 150 mm, determine (a) the common velocity after impact; (b) the average resistance of the ground. [4.85 m/s; (a) 2.83 m/s; (b) 14.70 kN]

22 Convert the following temperatures into the kelvin scale:
(a) 51°C; (b) −78°C; (c) 183°C. [(a) 324 K; (b) 195 K; (c) 456 K]

23 Convert the following temperatures into the Celsius scale:
(a) 307 K; (b) 237 K; (c) 415 K [(a) 34°C; (b) −36°C; (c) 142°C]

24 (a) What is the difference between heat and temperature?
(b) State three temperature-measuring devices and state the principle of operation of each.

25 Determine the quantity of heat energy (in megajoules) required to raise the temperature of 10 kg of water from 0°C to 50°C. Assume the specific heat capacity of water is 4200 J/(kg °C). [2.1 MJ]

26 Some copper, having a mass of 20 kg, cools from a temperature of 120°C to 70°C. If the specific heat capacity of copper is 390 J/(kg °C) how much heat energy is lost by the copper? [390 kJ]

27 A block of aluminium having a specific heat capacity of 950 J/(kg °C) is heated from 60°C to its melting point at 660°C. If the quantity of heat required is 2.85 MJ, determine the mass of the aluminium block. [5 kg]

28 20.8 kJ of heat energy is required to raise the temperature of 2 kg of lead from 16°C to 96°C. Determine the specific heat capacity of lead. [130 J/(kg °C)]

29 250 kJ of heat energy is supplied to 10 kg of iron which is initially at a temperature of 15°C. If the specific heat capacity of iron is 500 J/(kg °C), determine its final temperature. [65°C]

149

30 A brass container of mass 600 g contains 1.2 litres of water at 20°C. Determine the quantity of heat energy needed to raise the temperature of the water and its container to its boiling point assuming there are no heat losses. Assume that the specific heat capacity of brass is 370 J/(kg °C), the specific heat capacity of water is 4190 J/(kg °C) and one litre of water has a mass of 1 kg.   [420 kJ]

31 Some ice, initially at −40°C, has heat supplied to it at a constant rate until it becomes superheated steam at 150°C. Sketch a typical temperature–time graph expected and use it to explain the difference between sensible and latent heat.

32 How much heat is needed to completely melt 25 kg of ice at 0°C. Assume the specific latent heat of fusion of ice is 335 kJ/kg.   [8.375 MJ]

33 Determine the heat energy required to change 8 kg of water at 100°C to steam at 100°C. Assume the specific latent heat of vaporisation of water is 2260 kJ/kg.
[18.08 MJ]

34 Calculate the heat energy required to convert 10 kg of ice initially at −30°C completely into water at 0°C. Assume the specific heat capacity of ice is 2.1 kJ/(kg °C) and the specific latent heat of fusion of ice is 335 kJ/kg.   [3.98 MJ]

35 Determine the heat energy needed to convert completely 5 kg of water at 60°C to steam at 100°C, given that the specific heat capacity of water is 4.2 kJ/(kg °C) and the specific latent heat of vaporisation of water is 2260 kJ/kg.   [12.14 MJ]

36 Calculate how much heat energy is required to change 200 g of ice initially at a temperature of 243 K into superheated steam at 423 K. For ice, the specific heat capacity is 2.14 kJ/(kg K) and the latent heat of fusion is 335 kJ/kg. For water, the specific heat capacity is 4.19 kJ/(kg K) and the latent heat of vaporisation is 2260 kJ/kg. For steam, the specific heat capacity is 2.01 kJ/(kg K).   [635.7 kJ]

13 Thermal expansion

A. MAIN POINTS CONCERNED WITH THERMAL EXPANSION

1 When heat is applied to most materials, **expansion** occurs in all directions. Conversely, if heat energy is removed from a material (i.e. the material is cooled) **contraction** occurs in all directions.

   The effects of expansion and contraction each depend on the **change of temperature** of the material.

2 Some practical applications where expansion and contraction of solid materials must be allowed for include:

   (i) Overhead electrical transmission lines are hung so that they are slack in summer, otherwise their contraction in winter may snap the conductors or bring down pylons.

   (ii) Gaps need to be left in lengths of railway lines to prevent buckling in hot weather (except where these are continuously-welded).

   (iii) Ends of large bridges are often supported on rollers to allow them to expand and contract freely.

   (iv) Fitting a metal collar to a shaft or a steel tyre to a wheel is often achieved by first heating them so that they expand, fitting them in position, and then cooling them so that the contraction holds them firmly in place. This is known as a 'shrink-fit'. By a similar method hot rivets are used for joining metal sheets.

   (v) The amount of expansion varies with different materials. *Fig 1(a)* shows a bimetallic strip at room temperature, (i.e. two different strips of metal riveted together). When heated, brass expands more than steel, and since the two metals are riveted together the bimetallic strip is forced into an arc as shown in *Fig 1(b)*. Such a movement can be arranged to make or break an electric circuit and bimetallic strips are used, in particular, in thermostats (which are temperature operated switches) used to control central heating systems, cookers, refrigerators, toasters, irons, hot-water and alarm systems.

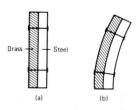

Fig 1

   (vi) Motor engines use the rapid expansion of heated gases to force a piston to move.

   (vii) Designers must predict, and allow for, the expansion of steel pipes in a steam-raising plant so as to avoid damage and consequent danger to health.

3   (i) Water is a liquid which at low temperatures displays an unusual effect. If cooled, contraction occurs until, at about 4°C, the volume is at a minimum. As the temperature is further decreased from 4°C to 0°C expansion occurs, i.e. the volume increases. When ice is formed considerable expansion occurs and it is this expansion which often causes frozen water pipes to burst.

(ii) A practical application of the expansion of a liquid is with thermometers, where the expansion of a liquid, such as mercury or alcohol, is used to measure temperature.

4   (i) The amount by which unit length of a material expands when the temperature is raised one degree is called the **coefficient of linear expansion** of the material and is represented by α (Greek alpha).

(ii) The units of the coefficient of linear expansion are m/(mK), although it is usually quoted as just /K or K^{-1}. For example, copper has a coefficient of linear expansion value of 17×10^{-6} K^{-1}, which means that a 1 m long bar of copper expands by 0.000 017 m if its temperature is increased by 1 K (or 1°C). If a 6 m long bar of copper is subjected to a temperature rise of 25 K then the bar will expand by $(6 \times 0.000\,017 \times 25)$ m, i.e. 0.002 55 m or 2.55 mm. (Since the kelvin scale uses the same temperature interval as the Celsius scale, a **change** of temperature of, say, 50°C, is the same as a change of temperature of 50 K.)

(iii) If a material, initially of length l_1 and at a temperature t_1 and having a coefficient of linear expansion α, has its temperature increased to t_2, then the new length l_2 of the material is given by:

New length = original length + expansion

i.e. $l_2 = l_1 + l_1\alpha(t_2 - t_1)$

i.e. $\boxed{l_2 = l_1[1 + \alpha(t_2 - t_1)]}$ (1)

(iv) Some typical values for the coefficient of linear expansion include:

Aluminium	23×10^{-6} K^{-1}	Brass	18×10^{-6} K^{-1}
Concrete	12×10^{-6} K^{-1}	Copper	17×10^{-6} K^{-1}
Gold	14×10^{-6} K^{-1}	Invar (nickel–	
Iron	11–12×10^{-6} K^{-1}	steel alloy)	0.9×10^{-6} K^{-1}
Steel	15–16×10^{-6} K^{-1}	Nylon	100×10^{-6} K^{-1}
Zinc	31×10^{-6} K^{-1}	Tungsten	4.5×10^{-6} K^{-1}

(See *Problems 1 to 4*)

5   (i) The amount by which unit area of a material increases when the temperature is raised by one degree is called the **coefficient of superficial (i.e. area) expansion** and is represented by β (Greek beta).

(ii) If a material having an initial surface area A_1 at temperature t_1 and having a coefficient of superficial expansion β, has its temperature increased to t_2, then the new surface area A_2 of the material is given by:

New surface area = original surface area + increase in area

i.e. $A_2 = A_1 + A_1\beta(t_2 - t_1)$

i.e. $\boxed{A_2 = A_1[1 + \beta(t_2 - t_1)]}$ (2)

(iii) It may be shown (see *Problem 5*) that the coefficient of superficial expansion is twice the coefficient of linear expansion, i.e. $\beta = 2\alpha$, to a very close approximation.

6   (i) The amount by which unit volume of a material increases for a one degree rise of temperature is called **the coefficient of cubic (or volumetric) expansion** and is represented by γ (Greek gamma).

(ii) If a material having an initial volume V_1 at temperature t_1 and having a coefficient of cubic expansion γ, has its temperature raised to t_2, then the new volume V_2 of the material is given by:

New volume = initial volume + increase in volume

i.e. $$V_2 = V_1 + V_1 \gamma (t_2 - t_1)$$

i.e. $$\boxed{V_2 = V_1 [1 + \gamma (t_2 - t_1)]} \tag{3}$$

(iii) It may be shown (see *Problem 6*) that the coefficient of cubic expansion is three times the coefficient of linear expansion, i.e. $\gamma = 3\alpha$, to a very close approximation. A liquid has no definite shape and only its cubic or volumetric expansion need be considered. Thus with expansions in liquids, equation (3) is used.

(iv) Some typical values for the coefficient of cubic expansion measured at $20°C$ (i.e. 293 K) include:

Ethyl alcohol	1.1×10^{-3} K^{-1}	Mercury	1.82×10^{-4} K^{-1}
Paraffin oil	9×10^{-2} K^{-1}	Water	2.1×10^{-4} K^{-1}

The coefficient of cubic expansion γ is only constant over a limited range of temperature.

(See *Problems 7 to 9*)

B. WORKED PROBLEMS ON THERMAL EXPANSION

Problem 1 The length of an iron steam pipe is 20.0 m long at a temperature of $18°C$. Determine the length of the pipe under working conditions when the temperature is $300°C$. Assume the coefficient of linear expansion of iron is 12×10^{-6} K^{-1}.

Length $l_1 = 20.0$ m; temperature $t_1 = 18°C$; $t_2 = 300°C$; $\alpha = 12 \times 10^{-6}$ K^{-1}
Length of pipe at $300°C$, $l_2 = l_1 [1 + \alpha(t_2 - t_1)]$

$$= 20.0 [1 + (12 \times 10^{-6})(300 - 18)]$$

$$= 20.0 [1 + 0.003\ 384] = 20.0 [1.003\ 384]$$

$$= 20.067\ 68 \text{ m}$$

i.e. an increase in length of 0.067 68 m, i.e. 67.68 mm.
In practice, allowances are made for such expansions. U-shaped expansion joints are connected into pilelines carrying hot fluids to allow some 'give' to take up the expansion.

Problem 2 An electrical overhead transmission line has a length of 80.0 m between its supports at $15°C$. Its length increases by 92 mm at $65°C$. Determine the coefficient of linear expansion of the material of the line.

Length $l_1 = 80.0$ m; $l_2 = 80.0$ m + 92 mm = 80.092 m; temperature $t_1 = 15°C$; temperature $t_2 = 65°C$.

Length $l_2 = l_1 [1 + \alpha(t_2 - t_1)]$, i.e. $80.092 = 80.0 [1 + \alpha(65 - 15)]$

$80.092 = 80.0 + (80.0)(\alpha)(50)$, i.e. $80.092 - 80.0 = (80.0)(\alpha)(50)$

Hence the coefficient of linear expansion, $\alpha = \dfrac{0.092}{(80.0)(50)} = 0.000\ 023$

i.e. $\alpha = 23 \times 10^{-6}$ K^{-1} (which is aluminium − see para. 4).

Problem 3 A measuring tape made of copper measures 5.0 m at a temperature of 288 K. Calculate the percentage error in measurement when the temperature has increased to 313 K. Take the coefficient of linear expansion of copper as 17×10^{-6} K^{-1}.

Length $l_1 = 5.0$ m; temperature $t_1 = 288$ K; $t_2 = 313$ K; $\alpha = 17 \times 10^{-6}$ K^{-1}
Length at 313 K, $l_2 = l_1[1 + \alpha(t_2 - t_1)] = 5.0[1 + (17 \times 10^{-6})(313 - 288)]$
$= 5.0[1 + (17 \times 10^{-6})(25)] = 5.0[1 + 0.000\,425]$
$= 5.0[1.000\,425] = 5.002\,125$ m,

i.e. the length of the tape has increased by 0.002 125 m.

Percentage error in measurement at 313 k $= \dfrac{\text{increase in length}}{\text{original length}} \times 100\%$

$= \dfrac{0.002\,125}{5.0} \times 100 = \mathbf{0.0425\%}$

Problem 4 The copper tubes in a boiler are 4.20 m long at a temperature of 20°C. Determine the length of the tubes (a) when surrounded only by feed water at 10°C; (b) when the boiler is operating and the mean temperature of the tubes is 320°C. Assume the coefficient of linear expansion of copper to be 17×10^{-6} K^{-1}.

(a) Initial length, $l_1 = 4.20$ m; initial temperature, $t_1 = 20$°C;
Final temperature, $t_2 = 10$°C; $\alpha = 17 \times 10^{-6}$ K^{-1}.
Final length at 10°C, $l_2 = l_1[1 + \alpha(t_2 - t_1)]$
$= 4.20[1 + (17 \times 10^{-6})(10 - 20)]$
$= 4.20[1 - 0.000\,17] = \mathbf{4.1993\ m}$
i.e. the tube contracts by 0.7 mm when the temperature decreases from 20°C to 10°C.

(b) $l_1 = 4.20$ m; $t_1 = 20$°C; $t_2 = 320$°C; $\alpha = 17 \times 10^{-6}$ K^{-1}
Final length at 320°C, $l_2 = l_1[1 + \alpha(t_2 - t_1)]$
$= 4.20[1 + (17 \times 10^{-6})(320 - 20)]$
$= 4.20[1 + 0.0051] = \mathbf{4.2214\ m,}$

i.e. the tube extends by 21.4 mm when the temperature rises from 20°C to 320°C.

Problem 5 Show that for a rectangular area of material having dimensions l by b the coefficient of superficial expansion $\beta \simeq 2\alpha$, where α is the coefficient of linear expansion.

Initial area, $A_1 = lb$. For a temperature rise of 1 K, side l will expand to $(l + l\alpha)$ and side b will expand to $(b + b\alpha)$. Hence the new area of the rectangle A_2 is given by

$A_2 = (l + l\alpha)(b + b\alpha) = l(1 + \alpha)b(1 + \alpha) = lb(1 + \alpha)^2$
$= lb(1 + 2\alpha + \alpha^2) \simeq lb(1 + 2\alpha)$, since α^2 is very small (see typical values in
para. 4(iv))

Hence $A_2 \simeq A_1(1 + 2\alpha)$

For a temperature rise of $(t_2 - t_1)$ K.

$A_2 \simeq A_1[1 + 2\alpha(t_2 - t_1)]$

Thus from equation (2), para. 4, $\beta \simeq 2\alpha$.

154

Initial volume, $V_1 = lbh$. For a temperature rise of 1 K, side l expands to $(l + l\alpha)$, side b expands to $(b + b\alpha)$ and side h expands to $(h + h\alpha)$. Hence the new volume of the block V_2 is given by:

$$V_2 = (l + l\alpha)(b + b\alpha)(h + h\alpha) = l(1 + \alpha)b(1 + \alpha)h(1 + \alpha)$$
$$= lbh(1 + \alpha)^3 = lbh(1 + 3\alpha + 3\alpha^2 + \alpha^3)$$
$$\simeq lbh(1 + 3\alpha), \text{ since terms in } \alpha^2 \text{ and } \alpha^3 \text{ are very small.}$$

Hence $V_2 \simeq V_1(1 + 3\alpha)$

For a temperature rise of $(t_2 - t_1)$ K,

$V_2 \simeq V_1[1 + 3\alpha(t_2 - t_1)]$

Thus from equation (3), para. 5, $\gamma \simeq 3\alpha$

(a) Initial diameter, $l_1 = 50$ mm; initial temperature, $t_1 = 289$ K; final temperature, $t_2 = 789$ K; $\alpha = 18 \times 10^{-6}$ K^{-1}

New diameter at 789 K, $l_2 = l_1[1 + \alpha(t_2 - t_1)]$, from equation (1),
$= 50[1 + (18 \times 10^{-6})(789 - 289)] = 50[1 + 0.009] = 50.45$ mm

Hence the increase in the diameter is **0.45 mm**

(b) Initial surface area of sphere, $A_1 = 4\pi r^2 = 4\pi\left(\dfrac{50}{2}\right)^2 = 2500\pi$ mm^2

New surface area at 789 K, $A_2 = A_1[1 + \beta(t_2 - t_1)]$, from equation (2),
$= A_1[1 + 2\alpha(t_2 - t_1)]$, since $\beta = 2\alpha$, to a very close approximation,
$= 2500\pi[1 + 2(18 \times 10^{-6})(500)] = 2500\pi[1 + 0.018] = 2500\pi + 2500\pi(0.018)$

Hence increase in surface area $= 2500\pi(0.018) =$ **141.4 mm^2**

(c) Initial volume of sphere, $V_1 = \frac{4}{3}\pi r^3 = \frac{4}{3}\pi\left(\dfrac{50}{2}\right)^3$ mm^3

New volume at 789 K, $V_2 = V_1[1 + \gamma(t_2 - t_1)]$, from equation (3)
$= V_1[1 + 3\alpha(t_2 - t_1)]$, since $\gamma = 3\alpha$, to a very close approximation
$= \frac{4}{3}\pi(25)^3[1 + 3(18 \times 10^{-6})(500)] = \frac{4}{3}\pi(25)^3[1 + 0.027]$
$= \frac{4}{3}\pi(25)^3 + \frac{4}{3}\pi(25)^3(0.027)$

Hence the increase in volume $= \frac{4}{3}\pi(25)^3(0.027) =$ **1767 mm^3**

Initial volume, $V_1 = 476$ mm^3; final volume $V_2 = 478$ mm^3; initial temperature $t_1 = 15$°C; $\gamma = 1.8 \times 10^{-4}$ K^{-1}.

Final volume, $V_2 = V_1[1 + \gamma(t_2 - t_1)]$, from equation (3),

$$V_2 = V_1 + V_1\gamma(t_2 - t_1)$$

from which, $(t_2 - t_1) = \dfrac{V_2 - V_1}{V_1 \gamma} = \dfrac{478 - 476}{(476)(1.8 \times 10^{-4})} = 23.34°C$

Hence $t_2 = 23.34 + 15 = 38.34°C$

Hence the temperature at which the volume of mercury is 478 mm^3 is **38.34°C**

Problem 9 A rectangular glass block has a length of 100 mm, width 50 mm and depth 20 mm at 293 K. When heated to 353 K its length increases by 0.054 mm. What is the coefficient of linear expansion of the glass?

Find also (a) the increase in surface area; (b) the change in volume resulting from the change of length.

Final length, $l_2 = l_1[1 + \alpha(t_2 - t_1)]$, from equation (1),

Hence increase in length is given by $l_2 - l_1 = l_1\alpha(t_2 - t_1)$

Hence $0.054 = (100)(\alpha)(353-293)$

from which the coefficient of linear expansion $\alpha = \dfrac{0.054}{(100)(60)} = 9 \times 10^{-6}$ K^{-1}

(a) Initial surface area of glass, $A_1 = (2 \times 100 \times 50) + (2 \times 50 \times 20) + (2 \times 100 \times 20)$
$= 10\,000 + 2000 + 4000 = 16\,000$ mm^2

Final surface area of glass, $A_2 = A_1[1 + \beta(t_2 - t_1)] = A_1[1 + 2\alpha(t_2 - t_1)]$, since $\beta = 2\alpha$ to a very close approximation

Hence increase in surface area $= A_1(2\alpha)(t_2 - t_1) = (16\,000)(2 \times 9 \times 10^{-6})(60)$
$= 17.28$ mm^2

(b) Initial volume of glass, $V_1 = 100 \times 50 \times 20 = 100\,000$ mm^3.

Final volume of glass, $V_2 = V_1[1 + \gamma(t_2 - t_1)] = V_1[1 + 3\alpha(t_2 - t_1)]$,

since $\gamma = 3\alpha$ to a very close approximation

Hence increase in volume of glass $= V_1(3\alpha)(t_2 - t_1)$
$= (100\,000)(3 \times 9 \times 10^{-6})(60) = 162$ mm^3

C. FURTHER PROBLEMS ON THERMAL EXPANSION

(a) SHORT ANSWER PROBLEMS

1 When heat is applied to most solids and liquids . . . . occurs.

2 When solids and liquids are cooled they usually . . . .

3 State three practical applications where the expansion of metals must be allowed for.

4 State a practical disadvantage where the expansions of metals occurs.

5 State one practical advantage of the expansion of liquids.

6 What is meant by the 'coefficient of expansion'.

7 Name the symbol and the unit used for the coefficient of linear expansion.

8 Define the 'coefficient of superficial expansion' and state its symbol.

9 Describe how water displays an unexpected effect between 0°C and 4°C.

10 Define the 'coefficient of cubic expansion' and state its symbol.

156

1   When the temperature of a rod of copper is increased, its length:
    (a) stays the same; (b) increases; (c) decreases.

2   The amount by which unit length of a material increases when the temperature is
    raised one degree is called the coefficient of:
    (a) cubic expansion; (b) superficial expansion; (c) linear expansion.

3   The symbol used for volumetric expansion is:
    (a) γ; (b) β; (c) l; (d) α.

4   A material of length l_1, at temperature θ_1 K is subjected to a temperature rise of
    θ K. The coefficient of linear expansion of the material is α K^{-1}. The material
    expands by:
    (a) $l_1(1 + \alpha\theta)$; (b) $l_1\alpha(\theta - \theta_1)$; (c) $l_1[1 + \alpha(\theta - \theta_1)]$; (d) $l_1\alpha\theta$

5   Some iron has a coefficient of linear expansion of 12×10^{-6} K^{-1}. A 100 mm
    length of iron piping is heated through 20 K. The pipe extends by:
    (a) 0.24 mm; (b) 0.024 mm; (c) 2.4 mm; (d) 0.0024 mm.

6   If the coefficient of linear expansion is A, the coefficient of superficial expansion
    is B and the coefficient of cubic expansion is C, which of the following is false?
    (a) $C = 3A$; (b) $A = \frac{B}{2}$; (c) $B = \frac{3}{2}C$; (d) $A = \frac{C}{3}$.

7   The length of a 100 mm bar of metal increases by 0.3 mm when subjected to a
    temperature rise of 100 K. The coefficient of linear expansion of the metal is:
    (a) 3×10^{-3} K^{-1}; (b) 3×10^{-4} K^{-1}; (c) 3×10^{-5} K^{-1}; (d) 3×10^{-6} K^{-1}.

8   A liquid has a volume V_1 at temperature θ_1. The temperature is increased to θ_2.
    If γ is the coefficient of cubic expansion, the increase in volume is given by:
    (a) $V_1\gamma(\theta_2 - \theta_1)$; (b) $V_1\gamma\theta_2$; (c) $V_1 + V_1\gamma\theta_2$; (d) $V_1[1 + \gamma(\theta_2 - \theta_1)]$

9   Which of the following statements is false?
    (a) Gaps need to be left in lengths of railway lines to prevent buckling in hot
        weather.
    (b) Bimetallic strips are used in thermostats, a thermostat having a temperature
        operated switch.
    (c) As the temperature of water is decreased from 4°C to 0°C contraction occurs.
    (d) A change of temperature of 15°C is equivalent to a change of temperature of
        15 K.

10 The volume of a rectangular block of iron at a temperature t_1 is V_1. The tempera-
    ture is raised to t_2 and the volume increases to V_2. If the coefficient of linear
    expansion of iron is α, then volume V_1 is given by:
    (a) $V_2[1 + \alpha(t_2 - t_1)]$; (b) $\dfrac{V_2}{1 + 3\alpha(t_2 - t_1)}$; (c) $3V_2\alpha(t_2 - t_1)$;
    (d) $\dfrac{1 + \alpha(t_2 - t_1)}{V_2}$.

(c) CONVENTIONAL PROBLEMS

1   A length of lead piping is 50.0 m long at a temperature of 16°C. When hot water
    flows through it the temperature of the pipe rises to 80°C. Determine the length
    of the hot pipe if the coefficient of linear expansion of lead is 29×10^{-6} K^{-1}.

[50.0928 m]

2   A rod of metal is measured at 285 K and is 3.521 m long. At 373 K the rod is
    3.523 m long. Determine the value of the coefficient of linear expansion for the
    metal.   [6.45×10^{-6} K^{-1}]

3   A copper overhead transmission line has a length of 40.0 m between its supports
    at 20°C. Determine the increase in length at 50°C if the coefficient of linear
    expansion of copper is 17×10^{-6} K^{-1}.   [20.4 mm]

4   A brass measuring tape measures 2.10 m at a temperature of 15°C. Determine
    (a) the increase in length when the temperature has increased to 40°C; (b) the
    percentage error in measurement at 40°C. Assume the coefficient of linear expansion
    of brass to be 18×10^{-6} K^{-1}.   [(a) 0.945 mm; (b) 0.045%]

5   A silver plate has an area of 800 mm^2 at 15°C. Determine the increase in the area
    of the plate when the temperature is raised to 100°C. Assume the coefficient of
    linear expansion of silver to be 19×10^{-6} K^{-1}.   [2.584 mm^2]

6   A pendulum of a 'grandfather' clock is 2.0 m long and made of steel. Determine
    the change in length of the pendulum if the temperature rises by 15 K. Assume the
    coefficient of linear expansion of steel to be 15×10^{-6} K^{-1}.   [0.45 mm]

7   A brass shaft is 15.02 mm in diameter and has to be inserted in a hole of diameter
    15.0 mm. Determine by how much the shaft must be cooled to make this possible,
    without using force. Take the coefficient of linear expansion of brass as
    18×10^{-6} K^{-1}.   [74 K]

8   A temperature control system is operated by the expansion of a zinc rod which is
    200 mm long at 15°C. If the system is set so that the source of heat supply is cut
    off when the rod has expanded by 0.20 mm, determine the temperature to which
    the system is limited. Assume the coefficient of linear expansion of zinc to be
    31×10^{-6} K^{-1}.   [47.26°C]

9   A brass collar of bore diameter 49.8 mm is to be shrunk on to a shaft of 50.0 mm
    diameter, both of these dimensions being measured at 20°C. To what temperature
    must the collar be heated so that it just slides on to the shaft. Assume the coefficient
    of linear expansion to be 18×10^{-6} K^{-1}.   [243.1°C]

10 A length of steel railway line is 30.0 m long when the temperature is 288 K.
    Determine the increase in length of the line when the temperature is raised to
    303 K. Assume the coefficient of linear expansion of steel to be 15×10^{-6} K^{-1}.
                                                                          [6.75 mm]

11 A steel girder is 8.0 m long at a temperature of 20°C. Determine the length of the
    girder at a temperature of 50°C. Take the coefficient of linear expansion for steel
    to be 15×10^{-6} K^{-1}.   [8.0036 m]

12 An aluminium ball is heated to a temperature of 773 K when it has a diameter of
    41.25 mm. It is then placed over a hole of diameter 41.0 mm. At what temperature
    will the ball just drop through the hole? Assume the coefficient of linear expansion
    of aluminium to be 23×10^{-6} K^{-1}.   [509.5 K]

13 At 283 K a thermometer contains 440 mm^3 of alcohol. Determine the temperature
    at which the volume is 480 mm^3 assuming that the coefficient of cubic expansion
    of the alcohol is 12×10^{-4} K^{-1}.   [358.8 K]

14 A zinc sphere has a radius of 30.0 mm at a temperature of 20°C. If the temperature
    of the sphere is raised to 420°C, determine the increase in: (a) the radius; (b) the

surface area; (c) the volume of the sphere. Assume the coefficient of linear expansion for zinc to be 31×10^{-6} K^{-1}.   [(a) 0.372 mm; (b) 280.5 mm^2; (c) 4207 mm^3]

15 A block of cast iron has dimensions of 50 mm by 30 mm by 10 mm at 15°C. Determine the increase in volume when the temperature of the block is raised to 75°C. Assume the coefficient of linear expansion of cast iron to be 11×10^{-6} K^{-1}.
[29.7 mm^3]

16 Two litres of water, initially at 20°C is heated to 40°C. Determine the volume of water at 40°C if the coefficient of volumetric expansion of water within this range is 30×10^{-5} K^{-1}.   [2.012 l]

17 Determine the increase in volume, in litres, of 3 m^3 of water when heated from 293 K to boiling point if the coefficient of cubic expansion is 2.1×10^{-4} K^{-1}. $(1 \ 1 \approx 10^{-3}$ m^3).   [50.4 l]

18 Determine the reduction in volume when the temperature of 0.5 l of ethyl alcohol is reduced from 40°C to -15°C. Take the coefficient of cubic expansion for ethyl alcohol as 1.1×10^{-3} K^{-1}.   [0.030 25 l]

14 Simple machines

A. MAIN POINTS CONCERNED WITH SIMPLE MACHINES

1   A machine is a device which can change the magnitude or line of action, or both magnitude and line of action of a force. A simple machine usually amplifies an input force, called the **effort**, to give a larger output force, called the **load**. Some typical examples of simple machines include pulley systems, screw-jacks, gear systems and lever systems.

2   The **force ratio** or **mechanical advantage** is defined as the ratio of load to effort, i.e.

$$\text{Force ratio} = \frac{\text{load}}{\text{effort}} \tag{1}$$

Since both load and effort are measured in newtons, force ratio is a ratio of the same units and thus is a dimensionless quantity.

3   The **movement ratio** or **velocity ratio** is defined as the ratio of the distance moved by the effort to the distance moved by the load, i.e.

$$\text{Movement ratio} = \frac{\text{distance moved by the effort}}{\text{distance moved by the load}} \tag{2}$$

Since the numerator and denominator are both measured in metres, movement ratio is a ratio of the same units and thus is a dimensionless quantity.

4   (i) The **efficiency of a simple machine** is defined as the ratio of the force ratio to the movement ratio, i.e.

$$\text{Efficiency} = \frac{\text{force ratio}}{\text{movement ratio}}$$

Since the numerator and denominator are both dimensionless quantities, efficiency is a dimensionless quantity. It is usually expressed as a percentage, thus:

$$\text{Efficiency} = \frac{\text{force ratio}}{\text{movement ratio}} \times 100 \text{ per cent} \tag{3}$$

(ii) Due to the effects of friction and inertia associated with the movement of any object, some of the input energy to a machine is converted into heat and losses occur. Since losses occur, the energy output of a machine is less than the energy input, thus the mechanical efficiency of any machine cannot reach 100%.

(iii) For simple machines, the relationship between effort and load is of the form: $F_e = aF_1 + b$, where F_e is the effort, F_1 is the load and a and b are constants. From equation (1),

$$\text{Force ratio} = \frac{\text{load}}{\text{effort}} = \frac{F_1}{F_e} = \frac{F_1}{aF_1 + b}$$

160

Dividing both numerator and denominator by F_1 gives:

$$\frac{F_1}{aF_1 + b} = \frac{1}{a + \dfrac{b}{F_1}},$$

When the load is large, F_1 is large and $\dfrac{b}{F_1}$ is small compared with a. The force ratio then becomes approximately equal to $1/a$ and is called the **limiting force ratio**. The **limiting efficiency** of a simple machine is defined as the ratio of the limiting force ratio to the movement ratio, i.e.

$$\text{Limiting efficiency} = \frac{1}{a \times \text{movement ratio}} \times 100 \text{ per cent},$$

where a is the constant for the law of the machine: $F_e = aF_1 + b$. Due to friction and inertia, the limiting efficiency of simple machines is usually well below 100%.

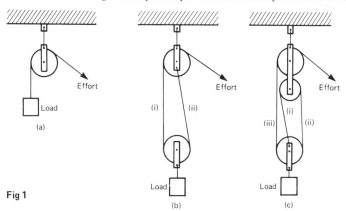

Fig 1

5 A **pulley system** is a simple machine.

A single-pulley system, shown in *Fig 1(a)*, changes the line of action of the effort, but does not change the magnitude of the force. A two-pulley system, shown in *Fig 1(b)*, changes both the line of action and the magnitude of the force. Theoretically, each of the ropes marked (i) and (ii) share the load equally, thus the theoretical effort is only half of the load, i.e. the theoretical force ratio is 2. In practice the actual force ratio is less than 2 due to losses. A three-pulley system is shown in *Fig 1(c)*. Each of the ropes marked (i), (ii) and (iii) carry one-third of the load, thus the theoretical force ratio is 3. In general, for a multiple pulley system having a total of n pulleys, the theoretical force ratio is n. Since the theoretical efficiency of a pulley system (neglecting losses) is 100% and since from equation (3)

$$\text{Efficiency} = \frac{\text{force ratio}}{\text{movement ratio}} \times 100 \text{ per cent, it follows that when the force ratio is } n,$$

$$100 = \frac{n}{\text{movement ratio}} \times 100,$$

that is, the movement ratio is also n.

6 A **simple screw-jack** is shown in *Fig 2* and is a simple machine since it changes both the magnitude and the line of action of a force. The screw of the table of the

jack is located in a fixed nut in the body of the jack. As the table is rotated by means of a bar, it raises or lowers a load placed on the table. For a single-start thread, as shown, for one complete revolution of the table, the effort moves through a distance $2\pi r$ and the load moves through a distance equal to the lead of the screw, say l. Thus:

$$\text{Movement ratio} = \frac{2\pi r}{l} \tag{4}$$

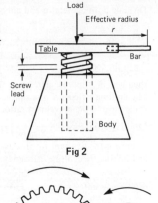

Fig 2

7 (i) A simple gear train is used to transmit rotary motion and can change both the magnitude and the line of action of a force, hence is a simple machine. The gear train shown in *Fig 3* consists of **spur gears** and has an effort applied to one gear, called the **driver** and a load applied to the other gear, called the **follower.**

(ii) In such a system, the teeth on the wheels are so spaced that they exactly fill the circumference with a whole number of identical teeth, and the teeth on the driver and follower mesh without interference. Under these conditions, the number of teeth on the driver and follower are in direct proportion to the circumference of these wheels, i.e.

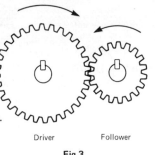

Driver   Follower

Fig 3

$$\frac{\text{Number of teeth on driver}}{\text{Number of teeth on follower}} = \frac{\text{circumference of driver}}{\text{circumference of follower}} \tag{5}$$

(iii) If there are, say, 40 teeth on the driver and 20 teeth on the follower then the follower makes two revolutions for each revolution of the driver. In general

$$\frac{\text{Number of revolutions made by driver}}{\text{Number of revolutions made by follower}} = \frac{\text{Number of teeth on follower}}{\text{Number of teeth on driver}} \tag{6}$$

It follows from equation (6) that the speeds of the wheels in a gear train are inversely proportional to the number of teeth.

(iv) The ratio of the speed of the driver wheel to that of the follower is the movement ratio, i.e.

$$\textbf{Movement ratio} = \frac{\textbf{speed of driver}}{\textbf{speed of follower}} = \frac{\textbf{teeth on follower}}{\textbf{teeth on driver}} \tag{7}$$

Fig 4

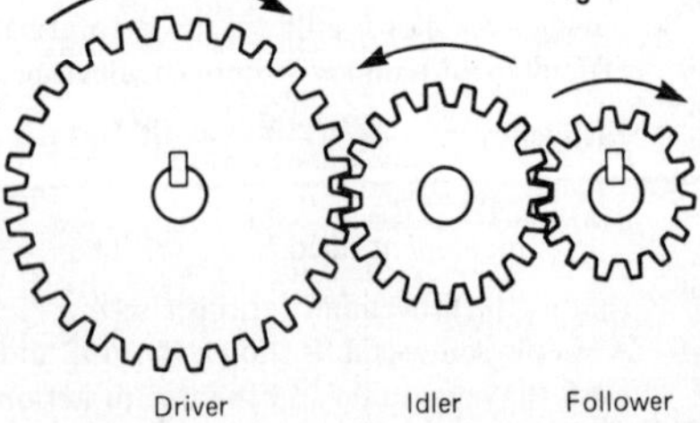

(v) When the same direction of rotation is required on both the driver and the follower an **idler wheel** is used as shown in *Fig 4*. Let the driver, idler and follower be A, B and C respectively, and let N be the speed of rotation and T be the number of teeth. Then from equation (7),

$$\frac{N_B}{N_A} = \frac{T_A}{T_B} \text{ and } \frac{N_C}{N_B} = \frac{T_B}{T_C}$$

Driver   Idler   Follower

Thus $\dfrac{\text{speed of A}}{\text{speed of C}} = \dfrac{N_A}{N_C} = \dfrac{N_B \dfrac{T_B}{T_A}}{N_B \dfrac{T_B}{T_C}} = \dfrac{T_B}{T_A} \times \dfrac{T_C}{T_B} = \dfrac{T_C}{T_A}$

This shows that the movement ratio is independent of the idler, only the direction of the follower being altered.

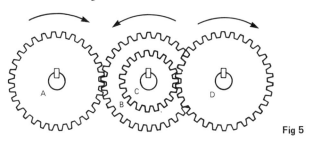

Fig 5

(iv) A compound gear train is shown in *Fig 5*, in which gear wheels B and C are fixed to the same shaft and hence $N_B = N_C$. From equation (7),

$\dfrac{N_A}{N_B} = \dfrac{T_B}{T_A}$, i.e. $N_B = N_A \times \dfrac{T_A}{T_B}$

Also, $\dfrac{N_D}{N_C} = \dfrac{T_C}{T_D}$, i.e., $N_D = N_C \times \dfrac{T_C}{T_D}$. But $N_B = N_C$,

hence $N_D = N_A \times \dfrac{T_A}{T_B} \times \dfrac{T_C}{T_D}$ (8)

For compound gear trains having, say, P gear wheels,

$N_P = N_A \times \dfrac{T_A}{T_B} \times \dfrac{T_C}{T_D} \times \dfrac{T_E}{T_F} \cdots \cdots \times \dfrac{T_O}{T_P}$ (9)

from which, **movement ratio** $= \dfrac{N_A}{N_P} = \dfrac{T_B}{T_A} \times \dfrac{T_D}{T_C} \cdots \cdots \times \dfrac{T_P}{T_O}$

8 A **lever** can alter both the magnitude and the line of action of a force and is thus classed as a simple machine. There are three types or orders of levers, as shown in *Fig 6*.

(i) A lever of the first order has the fulcrum placed between the effort and the load, as shown in *Fig 6(a)*.

(ii) A lever of the second order has the load placed between the effort and the fulcrum, as shown in *Fig 6(b)*

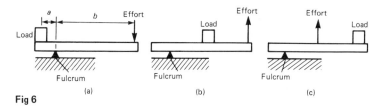

(a)   (b)   (c)

Fig 6

(iii) A lever of the third order has the effort applied between the load and the fulcrum, as shown in *Fig 6(c)*.

Problems on levers can largely be solved by applying the principle of moments (see Chapter 8). Thus for the lever shown in *Fig 6(a)*, when the lever is in equilibrium

Anticlockwise moment = clockwise moment

i.e. $a \times F_1 = b \times F_e$

Thus **force ratio** $= \dfrac{F_1}{F_e} = \dfrac{b}{a}$ (10)

$$= \frac{\text{distance of effort from fulcrum}}{\text{distance of load from fulcrum}}$$

B. WORKED PROBLEMS ON SIMPLE MACHINES

Problem 1 A simple machine raises a load of 160 kg through a distance of 1.6 m. The effort applied to the machine is 200 N and moves through a distance of 16 m. Taking g as 9.8 m/s², determine the force ratio, movement ratio and efficiency of the machine.

From equation (1), force ratio $= \dfrac{\text{load}}{\text{effort}} = \dfrac{160 \text{ kg}}{200 \text{ N}} = \dfrac{160 \times 9.8 \text{ N}}{200 \text{ N}} = \textbf{7.84}$

From equation (2), movement ratio $= \dfrac{\text{distance moved by effort}}{\text{distance moved by load}} = \dfrac{16 \text{ m}}{1.6 \text{ m}} = \textbf{10}$

From equation (3), efficiency $= \dfrac{\text{force ratio}}{\text{movement ratio}} \times 100$

$$= \frac{7.84}{10} \times 100 = \textbf{78.4\%}$$

Problem 2 For the simple machine of *Problem 1*, determine:

(a) The distance moved by the effort to move the load through a distance of 0.9 m;
(b) the effort which would be required to raise a load of 200 kg, assuming the same efficiency;
(c) the efficiency if, due to lubrication, the effort to raise the 160 kg load is reduced to 180 N.

(a) Since the movement ratio is 10, then from equation (2),

    Distance moved by the effort = 10 × distance moved by the load
                              = 10 × 0.9 = **9 m**

(b) Since the force ratio is 7.84, then from equation (1),

    Effort $= \dfrac{\text{load}}{7.84} = \dfrac{200 \times 9.8}{7.84} = \textbf{250 N}$

(c) The new force ratio $= \dfrac{\text{load}}{\text{effort}} = \dfrac{160 \times 9.8}{180} = 8.711$

    Hence new efficiency after lubrication $= \dfrac{8.711}{10} \times 100 = \textbf{87.11\%}$

Problem 3 In a test on a simple machine, the effort-load graph was a straight line of the form $F_e = aF_1 + b$. Two values lying on the graph were at $F_e = 10$ N; $F_1 = 30$ N; and at $F_e = 74$ N; $F_1 = 350$ N. The movement ratio of the machine was 17. Determine: (a) the limiting force ratio; (b) the limiting efficiency of the machine.

(a) The equation $F_e = aF_1 + b$ is of the form $y = mx + c$, where m is the gradient of the graph. The slope of the line passing through points (x_1, y_1) and (x_2, y_2) of the graph $y = mx + c$ is given by

$$m = \frac{y_2 - y_1}{x_2 - x_1} \ .$$

Thus for $F_e = aF_1 + b$, the slope a is given by

$$a = \frac{74 - 10}{350 - 30} = \frac{64}{320} = 0.2$$

But from para. 4(iii), the **limiting force ratio** is $\frac{1}{a}$, that is, **5**.

(b) The limiting efficiency $= \dfrac{1}{a \times \text{movement ratio}} \times 100$

$$= \frac{1}{0.2 \times 17} \times 100 = \mathbf{29.4\%}$$

Problem 4 A load of 80 kg is lifted by a three-pulley system similar to that shown in *Fig 1(c)*, page 161, and the applied effort is 392 N. Calculate (a) the force ratio; (b) the movement ratio; (c) the efficiency of the system. Take g to be $9.8 \, \text{m/s}^2$.

(a) From equation (1), the force ratio is given by $\dfrac{\text{load}}{\text{effort}}$. The load is 80 kg, i.e. 80×9.8 N.

Hence **force ratio** $= \dfrac{80 \times 9.8}{392} = \mathbf{2}$

(b) From para. 5, for a system having n pulleys, the movement ratio is n. Thus for a three-pulley system, the **movement ratio is 3**

(c) From equation (3), efficiency $= \dfrac{\text{force ratio}}{\text{movement ratio}} \times 100$

$$= \frac{2}{3} \times 100 = \mathbf{66.67\%}$$

Problem 5 A pulley system consists of two blocks, each containing three pulleys and connected as shown in *Fig 7*. An effort of 400 N is required to raise a load of 1500 N. Determine: (a) the force ratio; (b) the movement ratio; (c) the efficiency of the pulley system.

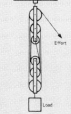

Fig 7

(a) From equation (1), force ratio $= \dfrac{\text{load}}{\text{effort}} = \dfrac{1500}{400} = 3.75$

(b) From para. 5, an n-pulley system has a movement ratio of n, hence this 6-pulley system has a **movement ratio of 6.**

(c) From equation (3), efficiency $= \dfrac{\text{force ratio}}{\text{movement ratio}} \times 100$

$$= \dfrac{3.75}{6} \times 100 = \textbf{62.5\%}$$

Problem 6 A screw-jack is being used to support the axle of a car, the load on it being 2.4 kN. The screw jack has an effort arm of effective radius 200 mm and a single-start square thread, having a lead of 5 mm. Determine the efficiency of the jack if an effort of 60 N is required to raise the car axle.

From equation (3), efficiency $= \dfrac{\text{force ratio}}{\text{movement ratio}} \times 100$ per cent

force ratio $= \dfrac{\text{load}}{\text{effort}} = \dfrac{2400 \text{ N}}{60 \text{ N}} = 40$

From equation (4), movement ratio $= \dfrac{2\pi r}{l} = \dfrac{2\pi \ 200 \text{ mm}}{5 \text{ mm}} = 251.3$

Hence efficiency $= \dfrac{40}{251.3} \times 100 = \textbf{15.9\%}$

Problem 7 A driver gear on the shaft of a motor has 35 teeth and meshes with a follower having 98 teeth. If the speed of the motor is 1400 revolutions per minute, find the speed of rotation of the follower.

From equation (7), $\dfrac{\text{speed of driver}}{\text{speed of follower}} = \dfrac{\text{teeth on follower}}{\text{teeth on driver}}$

i.e. $\dfrac{1400}{\text{speed of follower}} = \dfrac{98}{35}$

Hence, speed of follower $= \dfrac{1400 \times 35}{98} = \textbf{500 rev/min}$

Problem 8 A compound gear train similar to that shown in *Fig 5*, consists of a driver gear A, having 40 teeth, engaging with gear B, having 160 teeth. Attached to the same shaft as B, gear C has 48 teeth and meshes with gear D on the output shaft, having 96 teeth. Determine (a) the movement ratio of this gear system and (b) the efficiency when the force ratio is 6.

(a) From equation (8), the speed of D = speed of A $\times \dfrac{T_A}{T_B} \times \dfrac{T_C}{T_D}$

From equation (7), movement ratio $= \dfrac{\text{speed of A}}{\text{speed of D}} = \dfrac{T_B}{T_A} \times \dfrac{T_D}{T_C}$

$$= \dfrac{160}{40} \times \dfrac{96}{48} = 8$$

(b) The efficiency of any simple machine is $\dfrac{\text{force ratio}}{\text{movement ratio}} \times 100$ per cent

Thus, efficiency $= \dfrac{6}{8} \times 100 = \textbf{75\%}$

166

Problem 9 The load on a first-order lever, similar to that shown in *Fig 6(a)* on page 163, is 1.2 kN. Determine the effort, the force ratio, and the movement ratio when the distance between the fulcrum and load is 0.5 m and the distance between the fulcrum and effort is 1.5 m. Assume the lever is 100% efficient.

Applying the principle of moments, for equilibrium:

Anticlockwise moment = clockwise moment

i.e. 1200 N × 0.5 m = effort × 1.5 m

Hence, effort $= \dfrac{1200 \times 0.5}{1.5} =$ **400 N**

From para. 8, force ratio $= \dfrac{F_1}{F_e} = \dfrac{1200}{400} = 3$

Alternatively, force ratio $= \dfrac{b}{a} = \dfrac{1.5}{0.5} = 3$

This result shows that to lift a load of say 300 N, an effort of 100 N is required.

Since, from equation (3), efficiency $= \dfrac{\text{force ratio}}{\text{movement ratio}} \times 100$ per cent,

then movement ratio $= \dfrac{\text{force ratio}}{\text{efficiency}} \times 100 = \dfrac{3}{100} \times 100 = 3$

This result shows that to raise the load by, say, 100 mm, the effort has to move 300 mm.

Problem 10 A second-order lever, AB, is in a horizontal position. The fulcrum is at point C. An effort of 60 N applied at B just moves a load at point D, when BD is 0.5 m and BC is 1.25 m. Calculate the load and the force ratio of the lever.

A second-order lever system is shown in *Fig 6(b)* on page 163. Taking moments about the fulcrum as the load is just moving, gives:

Anticlockwise moments = clockwise moments

i.e. 60 N × 1.25 m = load × 0.75 m

Thus, load $= \dfrac{60 \times 1.25}{0.75} = 100$ N

From equation (1), force ratio $= \dfrac{\text{load}}{\text{effort}} = \dfrac{100}{60} = 1\tfrac{2}{3}$

Alternatively, from para 8, force ratio $= \dfrac{\text{distance of effort from fulcrum}}{\text{distance of load from fulcrum}}$

$$= \dfrac{1.25}{0.75} = 1\tfrac{2}{3}$$

C. FURTHER PROBLEMS ON SIMPLE MACHINES

(a) SHORT ANSWER PROBLEMS

1 State what is meant by a simple machine.

2 Define force ratio.

3 Define movement ratio.

4 Define the efficiency of a simple machine in terms of the force and movement ratios.

5 State briefly why the efficiency of a simple machine cannot reach 100%.

6 With reference to the law of a simple machine, state briefly what is meant by the term 'limiting force ratio'.

7 Define limiting efficiency.

8 Explain why a four-pulley system has a force ratio of 4 when losses are ignored.

9 Give the movement ratio for a screw-jack in terms of the effective radius of the effort and the screw lead.

10 Explain the action of an idler gear.

11 Define the movement ratio for a two-gear system in terms of the teeth on the wheels.

12 Show that the action of an idler wheel does not affect the movement ratio of a gear system.

13 State the relationship between the speed of first gear and the speed of the last gear in a compound train of four gears, in terms of the teeth on the wheels.

14 Sketch a second-order lever system.

15 Define the force ratio of a first-order lever system in terms of the distances of the load and effort from the fulcrum.

(b) MULTI-CHOICE PROBLEMS (answers on page 173)

A simple machine requires an effort of 250 N moving through 10 m to raise a load of 1000 N through 2 m. Use this data to find the correct answers to *Problems 1 to 3*, selecting these answers from:
(a) $\frac{1}{4}$; (b) 4; (c) 80%; (d) 20%; (e) 100; (f) 5; (g) 100%; (h) $\frac{1}{5}$; (i) 25%.

1 Find the force ratio.

2 Find the movement ratio.

3 Find the efficiency.

The law of a machine is of the form $F_e = aF_1 + b$. An effort of 12 N is required to raise a load of 40 N and an effort of 6 N is required to raise a load of 16 N. The movement ratio of the machine is 5. Use this data to find the correct answers to *Problems 4 to 6*, selecting these answers from:
(a) 80%; (b) 4; (c) 2.8; (d) $\frac{1}{4}$; (d) $\frac{1}{2.8}$; (f) 25%; (g) 100%; (h) 2; (i) 25%.

4 Determine the constant 'a'.
5 Find the limiting force ratio.
6 Find the limiting efficiency.

7 Which of the following statements is false?
   (a) A single-pulley system changes the line of action of the force but does not change the magnitude of the force, when losses are neglected.
   (b) In a two-pulley system, the force ratio is $\frac{1}{2}$ when losses are neglected.
   (c) In a two-pulley system, the movement ratio is 2.
   (d) The efficiency of a two-pulley system is 100% when losses are neglected.

168

8 Which of the following statements concerning a screw-jack is false?
   (a) A screw-jack changes both the line of action and the magnitude of the force.
   (b) For a single-start thread, the distance moved in 5 revolutions of the table is 5 l, where l is the lead of the screw.
   (c) The distance moved by the effort is $2\pi r$, where r is the effective radius of the effort.
   (d) The movement ratio is given by $\dfrac{2\pi r}{5\,l}$

9 In a simple gear train, a follower has 50 teeth and the driver has 30 teeth. The movement ratio is: (a) $\frac{3}{5}$; (b) 20; (c) $\frac{5}{3}$; (d) 80.

10 Which of the following statements is true?
   (a) An idler wheel between a driver and a follower is used to make the direction of the follower opposite to that of the driver.
   (b) An idler wheel is used to change the movement ratio.
   (c) An idler wheel is used to change the force ratio.
   (d) An idler wheel is used to make the direction of the follower the same as that of the driver.

11 Which of the following statements is false?
   (a) In a first-order lever, the fulcrum is between the load and the effort.
   (b) In a second-order lever, the load is between the effort and the fulcrum.
   (c) In a third-order lever, the effort is applied between the load and the fulcrum.
   (d) The force ratio for a first-order lever system is given by:
       distance of load from fulcrum
       distance of effort from fulcrum

12 In a second-order lever system, the load is 200 mm and the effort is 500 mm from the fulcrum. If losses are neglected, an effort of 100 N will raise a load of:
   (a) 100 N; (b) 250 N; (c) 400 N; (d) 40 N.

(c) CONVENTIONAL PROBLEMS

1 A simple machine raises a load of 825 N through a distance of 0.3 m. The effort is 250 N and moves through a distance of 3.3 m. Determine: (a) the force ratio; (b) the movement ratio; (c) the efficiency of the machine at this load.
[(a) 3.3; (b) 11; (c) 30%]

2 The efficiency of a simple machine is 50%. If a load of 1.2 kN is raised by an effort of 300 N, determine the movement ratio. [8]

3 An effort of 10 N applied to a simple machine moves a load of 40 N through a distance of 100 mm, the efficiency at this load being 80%. Calculate: (a) the movement ratio; (b) the distance moved by the effort. [(a) 5; (b) 500 mm]

4 The effort required to raise a load using a simple machine, for various values of load is as shown.

| Load (N) | 2050 | 4120 | 7410 | 8240 | 10 300 |
| Effort (N) | 260 | 340 | 435 | 505 | 580 |

If the movement ratio for the machine is 30, determine: (a) the law of the machine; (b) the limiting value of force ratio; (c) the limiting efficiency.
[(a) $F_e = 0.4\,F_1 + 170$; (b) 25; (c) $83\frac{1}{3}\%$]

5 For the data given in *Problem 4*, determine the values of force ratio and efficiency for each value of the load. Hence plot graphs of effort, force ratio and efficiency to

a base of load. From the graphs, determine the effort required to raise a load of 6 kN and the efficiency at this load.   [410 N; 48.8%]

6 In an experiment involving a screw-jack, the following data was obtained.

Load (N)	0	250	500	750	1000	1250
Effort (N)	10	30	50	70	90	110

Determine the law of the machine. If the movement ratio is 45, find the limiting force ratio and the limiting efficiency. $[F_e = 0.08 F_1 + 10; 12.5; 27.8\%]$

7 For the data given in Problem 6, calculate the force ratio and efficiency for each value of the loads. Plot graphs of effort, force ratio and efficiency to a base of load and hence determine the effort required to raise a load of 850 N. Determine the efficiency of the screw-jack at this load.   [78 N; 24.2%]

8 A pulley system consists of four pulleys in an upper block and three pulleys in a lower block. Make a sketch of this arrangement showing how a movement ratio of 7 may be obtained. If the force ratio is 4.2, what is the efficiency of the pulley system?   [60%]

9 A three-pulley lifting system is used to raise a load of 4.5 kN. Determine the effort required to raise this load when losses are neglected. If the actual effort required is 1.6 kN, determine the efficiency of the pulley system at this load.
                    [1.5 kN; 93.75%]

10 Sketch a simple screw-jack. The single-start screw of such a jack has a lead of 6 mm and the effective length of the operating bar from the centre of the screw is 300 mm. Calculate the load which can be raised by an effort of 150 N if the efficiency at this load is 20%.   [9.425 kN]

11 A load of 1.7 kN is lifted by a screw-jack having a single-start screw of lead 5 mm. The effort is applied at the end of an arm of effective length 300 mm from the centre of the screw. Calculate the effort required if the efficiency at this load is 20%.
                    [22.55 N]

12 The driver gear of a gear system has 28 teeth and meshes with a follower gear having 168 teeth. Determine the movement ratio and the speed of the follower when the driver gear rotates at 60 revolutions per second.   [6; 10 rev/s]

13 A compound gear train has a 30-tooth driver gear, A, meshing with a 90-tooth follower gear B. Mounted on the same shaft as B and attached to it is a gear C with 60 teeth, meshing with a gear D on the output shaft having 120 teeth. Calculate the movement and force ratios if the overall efficiency of the gears is 72%.
                    [6; 4.32]

14 A compound gear train is as shown in Fig 5 on page 163. The movement ratio is 6 and the numbers of teeth on gears A, C and D are 25, 100 and 60 respectively. Determine the number of teeth on gear B and the force ratio when the efficiency is 60%.   [250; 3.6]

15 Use sketches to show what is meant by: (a) a first-order; (b) a second-order; (c) a third-order lever system. Give one practical use for each type of lever.

16 In a second-order lever system, the force ratio is 2.5. If the load is at a distance of 0.5 m from the fulcrum, find the distance that the effort acts from the fulcrum if losses are negligible.   [1.25 m]

17 A lever AB is 2 m long and the fulcrum is at a point 0.5 m from B. Find the effort to be applied at A to raise a load of 0.75 kN at B when losses are negligible.

[250 N]

18 The load on a third-order lever system is at a distance of 750 mm from the fulcrum and the effort required to just move the load is 1 kN when applied at a distance of 250 mm from the fulcrum. Determine the value of the load and the force ratio, if losses are negligible. $[333\frac{1}{3} \text{ N}, \frac{1}{3}]$

Answers to multi-choice problems

Chapter 1 (page 13)

1 (b); 2 (c); 3 (d); 4 (d); 5 (b); 6 (b); 7 (c); 8 (a); 9 (c); 10 (b); 11 (d); 12 (b); 13 (a); 14 (c); 15 (a).

Chapter 2 (page 27)

1 (d); 2 (b); 3 (d); 4 (c); 5 (d); 6 (a); 7 (b); 8 (c).

Chapter 3 (page 39)

1 (c); 2 (b); 3 (c); 4 (b); 5 (c); 6 (a); 7 (c); 8 (b); 9 (a); 10 (b) and (c).

Chapter 4 (page 50)

1 (c); 2 (d); 3 (d); 4 (a); 5 (d); 6 (c); 7 (b); 8 (c); 9 (b).

Chapter 5 (page 66)

1 (d); 2 (a) or (c); 3 (b); 4 (b); 5 (c); 6 (f); 7 (c); 8 (a); 9 (i); 10 (j); 11 (g); 12 (c); 13 (b); 14 (p); 15 (d); 16 (o); 17 (n).

Chapter 6 (page 84)

1 (b); 2 (c); 3 (c); 4 (b); 5 (d); 6 (b); 7 (c); 8 (f); 9 (h); 10 (d).

Chapter 7 (page 97)

1 (b); 2 (b); 3 (c); 4 (d); 5 (c); 6 (d).

Chapter 8 (page 107)

1 (a); 2 (c); 3 (a); 4 (d); 5 (a); 6 (d); 7 (c); 8 (a); 9 (d); 10 (c).

Chapter 9 (page 117)

1 (c); 2 (a); 3 (d); 4 (c); 5 (b); 6 (d); 7 (g); 8 (i); 9 (c); 10 (k).

Chapter 10 (page 124)

1 (b); 2 (b); 3 (a); 4 (a); 5 (b); 6 (c); 7 (d); 8 (d); 9 (c); 10 (b).

Chapter 11 (page 130)

1 (f); 2 (e); 3 (i); 4 (c); 5 (h); 6 (b); 7 (d); 8 (a).

Chapter 12 (page 147)

1 (c); 2 (c); 3 (a); 4 (d); 5 (c); 6 (b); 7 (d); 8 (b); 9 (a); 10 (c); 11 (b); 12 (b).

Chapter 13 (page 157)

1 (b); 2 (c); 3 (a); 4 (d); 5 (b); 6 (c); 7 (c); 8 (a); 9 (c); 10 (b).

Chapter 14 (page 168)

1 (b); 2 (f); 3 (c); 4 (d); 5 (b); 6 (a); 7 (b); 8 (d); 9 (c); 10 (d); 11 (d); 12 (b).

Index

Mathematics

Mathematics for Technicians 1
F Tabberer

1978   192 pages   246 × 189 mm
0 408 00326 X   Limp   Illustrated

Mathematics for Technicians 2
F Tabberer

1978   156 pages   246 × 189 mm
0 408 00371 5   Limp   Illustrated

Science

Physical Science for Technicians 1
R McMullan

1978   96 pages   246 × 189 mm
0 408 00332 4   Limp   Illustrated

Building Construction, Civil Engineering, Surveying and Architecture

Building Technology 1
J T Bowyer

1978   96 pages   246 × 189 mm
0 408 00298 0   Limp   Illustrated

Building Technology 2
J T Bowyer

1978   96 pages   246 × 189 mm
0 408 00299 9   Limp   Illustrated

Building Technology 3
J T Bowyer

1980   104 pages   246 × 189 mm
0 408 00411 8   Limp   Illustrated

Civil Engineering Technology 3
B J Fletcher and S A Lavan

1980   96 pages   246 × 189 mm
0 408 00426 6   Limp   Illustrated

•Construction Science and Materials 2
D Watkins and J Fincham

1981   192 pages approx   246 × 189 mm
0 408 00488 6   Limp   Illustrated

•Site Surveying and Levelling 2
W S Whyte and R E Paul

1981   160 pages approx   246 × 189 mm
0 408 00532 7   Limp   Illustrated

Heating and Hot Water Services for Technicians
K Moss

1978   168 pages   246 × 189 mm
0 408 00300 6   Limp   Illustrated

Electrical, Electronic and Telecommunications Engineering

Electrical Drawing for Technicians 1
F Linsley

1979   96 pages   246 × 189 mm
0 408 00417 7   Limp   Illustrated

Telecommunications Systems for Technicians 1
G L Danielson and R S Walker

1979   112 pages   246 × 189 mm
0 408 00352 9   Limp   Illustrated

•Transmission Systems for Technicians 2
G L Danielson and R S Walker

1981   72 pages approx   246 × 189 mm
0 408 00562 9   Limp   Illustrated

•Radio Systems for Technicians 2
G L Danielson and R S Walker

1981   96 pages approx   246 × 189 mm
0 408 00561 0   Limp   Illustrated

•Radio Systems for Technicians 3
G L Danielson and R S Walker

1982   112 pages approx   246 × 189 mm
0 408 00588 2   Limp   Illustrated

Electrical and Electronic Principles 2
I R Sinclair

1979   96 pages   246 × 189 mm
0 408 00433 9   Limp   Illustrated

Electrical and Electronic Applications 2
D W Tyler

1980   204 pages   246 × 189 mm
0 408 00412 6   Limp   Illustrated

Electronics for Technicians 2
S A Knight

1978 112 pages 246 × 189 mm
0 408 00324 3 Limp Illustrated

Electronics for Technicians 3
S A Knight

1980 160 pages 246 × 189 mm
0 408 00458 4 Limp Illustrated

Electrical Principles for Technicians 2
S A Knight

1978 144 pages 246 × 189 mm
0 408 00325 1 Limp Illustrated

Electrical and Electronic Principles 3
S A Knight

1980 160 pages 246 × 189 mm
0 408 00456 8 Limp Illustrated

•Electrical and Electronic Principles 4/5
S A Knight

1982 176 pages approx 246 × 189 mm
0 408 01109 2 Limp Illustrated

Mechanical, Production, Marine and Motor Vehicle Engineering

Vehicle Technology 1
M J Nunney

1980 112 pages 246 × 189 mm
0 408 00461 4 Limp Illustrated

•Vehicle Technology 2
M J Nunney

1981 96 pages approx 246 × 189 mm
0 408 00594 7 Limp Illustrated

•Engine Technology 1
M J Nunney

1981 120 pages approx 246 × 189 mm
0 408 00511 4 Limp Illustrated

Manufacturing Technology 2
P J Harris

1979 96 pages 246 × 189 mm
0 408 00410 X Limp Illustrated

•Manufacturing Technology 3
P J Harris

1981 104 pages approx 246 × 189 mm
0 408 00493 2 Limp Illustrated

•Fabrication, Welding and Metal Joining Processes — A textbook for Technicians and Craftsmen
C Flood

1981 160 pages approx 246 × 189 mm
0 408 00448 7 Limp Illustrated

•Materials Technology for Technicians 2
W Bolton

1981 128 pages approx 246 × 189 mm
0 408 01117 3 Limp Illustrated

•Materials Technology for Technicians 3
W Bolton

1982 128 pages approx 246 × 189 mm
0 408 01116 5 Limp Illustrated

•Materials Technology 4
W Bolton

1981 128 pages approx 246 × 189 mm
0 408 00584 X Limp Illustrated

Mechanical Science for Technicians 3
W Bolton

1980 128 pages 246 × 189 mm
0 408 00486 X Limp Illustrated

•Mechanical Science for Higher Technicians 4/5
D H Bacon and R C Stephens

1981 256 pages approx 234 × 156 mm
0 408 00570 X Limp Illustrated

•Thermodynamics for Technicians 3/4
D H Bacon and R C Stephens

1982 96 pages approx 234 × 156 mm
0 408 01114 9 Limp Illustrated

Engineering Instrumentation and Control
W Bolton

1980 144 pages 246 × 189 mm
0 408 00462 2 Limp Illustrated

Butterworths
Checkbooks

Checkbook General Editors

for | **Mathematics, Sciences, Electrical, Electronic, Telecommunications, Mechanical, Production, Marine and Motor Vehicle Engineering**

John O Bird and Anthony J C May of Highbury College of Technology, Portsmouth.

for | **Building Construction, Civil Engineering, Surveying and Architecture**

Colin R Bassett, lately of Guildford County College of Technology.

Mathematics

•Mathematics 1 Checkbook

J O Bird and A J C May

1981 168 pages approx 186 × 123 mm
0 408 00609 9 Limp Illustrated
0 408 00632 3 Cased

•Mathematics 2 Checkbook

J O Bird and A J C May

1981 272 pages approx 186 × 123 mm
0 408 00610 2 Limp Illustrated
0 408 00633 1 Cased

•Mathematics 3 Checkbook

J O Bird and A J C May

1981 196 pages approx 186 × 123 mm
0 408 00611 0 Limp Illustrated
0 408 00634 X Cased

•Mathematics 4 Checkbook

J O Bird and A J C May

1981 240 pages approx 186 × 123 mm
0 408 00612 9 Limp Illustrated
0 408 00660 9 Cased

•Engineering Mathematics and Science 3 Checkbook

J O Bird, A J C May and D Ayling

1982 128 pages approx 186 × 123 mm
0 408 00625 0 Limp Illustrated

Sciences

•Chemistry 2 Checkbook

P J Chivers

1981 144 pages approx 186 × 123 mm
0 408 00622 6 Limp Illustrated
0 408 00637 4 Cased

•Chemistry 3 Checkbook

P J Chivers

1982 144 pages approx 186 × 123 mm
0 408 00658 7 Limp Illustrated
0 408 00662 5 Cased

•Building Science and Materials 2 Checkbook
and
Environmental Science 3/4 Checkbook
(see under Building Construction, etc. below)

- **Electrical Science 3 Checkbook**
  (see under Electrical, Electronics, etc. below)

- **Engineering Mathematics and Science 3 Checkbook**
  (see under Mathematics above)

- **Mechanical Science 3 Checkbook**
  (see under Mechanical, Production, etc. below)

Building Construction, Civil Engineering, Surveying and Architecture

- **Construction Drawing 1 Checkbook**

  ### J Greening and A Bowers

  1982   144 pages approx   186 × 123 mm
  0 408 00646 3   Limp   Illustrated

- **Construction Technology 1 Checkbook**

  ### R Chudley

  1981   160 pages approx   186 × 123 mm
  0 408 00602 1   Limp   Illustrated
  0 408 00642 0   Cased

- **Construction Technology 2 Checkbook**

  ### R Chudley

  1981   160 pages approx   186 × 123 mm
  0 408 00603 X   Limp   Illustrated
  0 408 00664 1   Cased

- **Building Science and Materials 2 Checkbook**

  ### M D W Pritchard

  1981 116 pages approx   186 × 123 mm
  0 408 00607 2   Limp   Illustrated
  0 408 00640 4   Cased

- **Environmental Science 3/4 Checkbook**

  ### M D W Pritchard

  1982   144 pages approx   186 × 123 mm
  0 408 00608 0   Limp   Illustrated
  0 408 00663 3   Cased

- **Geotechnics 4 Checkbook**

  ### R Whitlow

  1982   144 pages approx   186 × 123 mm
  0 408 00631 5   Limp   Illustrated

- **Building Law 4 Checkbook**

  ### A Galbraith

  1982   112 pages approx   186 × 123 mm
  0 408 00588 1   Limp   Illustrated

- **Economics for the Construction Industry 4 Checkbook**

  ### V J Seddon and G B Atkinson

  1982   144 pages approx   186 × 123 mm
  0 408 00655 2   Limp   Illustrated

- **Building Services and Equipment 4 Checkbook**

  ### F Hall

  1981   144 pages approx   186 × 123 mm
  0 408 00613 7   Limp   Illustrated
  0 408 00641 2   Cased

- **Building Services and Equipment 5 Checkbook**

  ### F Hall

  1981   144 pages approx   186 × 123 mm
  0 408 00614 5   Limp   Illustrated
  0 408 00651 X   Cased

Electrical, Electronic and Telecommunications Engineering

- **Microelectronic Systems 1 Checkbook**

  ### R Vears

  1981   84 pages approx   186 × 123 mm
  0 408 00552 1   Limp   Illustrated
  0 408 00638 2   Cased

- **Microelectronic Systems 2 Checkbook**

  ### R Vears

  1982   160 pages approx   186 × 123 mm
  0 408 00659 5   Limp   Illustrated